ANIMAL SCIENCE, ISSUES AND RESEARCH

ANIMAL COMMUNICATION AND COGNITION

PRINCIPLES, EVOLUTION AND DEVELOPMENT

ANIMAL SCIENCE, ISSUES AND RESEARCH

Additional books in this series can be found on Nova's website under the Series tab.

Additional e-books in this series can be found on Nova's website under the e-book tab.

ANIMAL SCIENCE, ISSUES AND RESEARCH

ANIMAL COMMUNICATION AND COGNITION

PRINCIPLES, EVOLUTION AND DEVELOPMENT

TABITHA WAGNER
EDITOR

nova publishers
New York

Library of Congress Cataloging-in-Publication Data

Animal communication and cognition : principles, evolution, and development / editor, Tabitha Wagner.
pages cm. -- (Animal science, issues and research)
Includes index.
ISBN 978-1-63482-411-8 (hardcover)
1. Animal communication. I. Wagner, Tabitha.
QL776.A533 2015
591.59--dc23

2015007250

Published by Nova Science Publishers, Inc. † New York

CONTENTS

Preface		**vii**
Chapter 1	Computer-Animated Stimuli to Investigate Lizard Visual Communication: A Case Study in the Jacky Dragon *Kevin L. Woo and Guillaume Rieucau*	**1**
Chapter 2	Pretend Play in Signing Chimpanzees (Pan troglogytes) *Tennyson E. Egan and Mary Lee A. Jensvold*	**57**
Chapter 3	Evolution of TRP Channels As Mediators of Pheromone Signaling *Rory Ostrow, Travers Ruel, Tommy Voigt, Cecil James Saunders and Robert Hallock*	**87**
Chapter 4	The Role of Cuticular Hydrocarbon-Based Chemical Communication in Prezygotic Reproductive Isolation *J. Buellesbach and T. Schmitt*	**107**
Index		**121**

PREFACE

This book begins with a discussion on computer-animated stimuli to investigate visual communication in lizards. It then discusses the pretend play in signing chimpanzees; the evolution of TRP channels as mediators of pheromone signaling; and the role of cuticular hydrocarbon-based chemical communication in prezygotic reproductive isolation.

Chapter 1 – Visual signals that are designed for effective communication are complex. Over the years, researchers interested in the study of animal communication have employed various artificial stimuli during behavioral experiments to precisely control what the observers get to see or experience. These diverse techniques to stage animal interactions have offered an alternative to the traditional experimental methods relying on the use of live companions or demonstrators. Current technological advances allow researchers to develop realistic computer-animated stimuli of social partners with high degree of fidelity for both morphological and dynamic behavioral characteristics. Because the computer-generated animation technique offers researchers with the ability to produce standardized visual stimuli while reducing the variable behavior of demonstrators across experimental sessions, its used is becoming increasingly popular in behavioral experiments in various taxa. In the following chapter, we provide an appraisal of the computer-generated animations efficiency to stage social interactions in lizards in general and the Jacky dragon (*Amphibolurus muricatus*) in particular. The authors present relevant experimental studies in which this innovative technique has been employed to simulate social partners in the Jacky dragon to address diverse ecological important questions. In addition, they consider the future of video and animation playback in the study of lizard's communication.

Chapter 2 – Pretend play is well studied in human children and is thought to be associated with symbolic thought and theory of mind. There are many descriptions of pretend play in chimpanzees and other apes but few systematic studies. This study was a systematic sampling of videotapes for instances of pretend play in chimpanzees who have acquired American Sign Language. There were two sampling methods. The first sample of 17 hours 36 minutes and 26 seconds of videotape contained five instances of pretend play. The second sample of 50 hours 16 minutes and 41 seconds contained 16 instances. The instances fell into four different categories; substitution, attribution of function, animation, and insubstantial situation attribution.

Chapter 3 – Members of the transient receptor potential (TRP) channel family are involved in pheromone signaling, in both the vomeronasal and the olfactory systems in many species. Since the light-activated TRP and TRPL channels were discovered in *Drosophila*, TRP channel homologues have been identified in many different sensory systems across taxa. For example TRPC2, TRPM4 and TRPM5 are essential for vomeronasal and olfactory system sensing in many animals. Although these channels serve a very different function from their *Drosophila* homologues, their mechanism of action is similar to most TRP channel members utilized in the mammalian chemical senses--activation of a g-protein coupled receptor triggers opening of the TRP channel through the production of a secondary messenger, resulting the influx of cations and the depolarization of the sensory cell. Since, pheromone signaling plays a crucial role in the communication of many diverse non-human animals and TRP channel family members are essential for these systems to function, examining the difference between TRPC2, TRPM4 and TRPM5 may reveal how evolution has shaped these sensory systems. This review describes the array of TRP channels important to pheromone sensing, the evolution processes which shaped them, and their involvement in the transduction of pheromone signals.

Chapter 4 – One of the most important questions in evolutionary biology is how new species originate. In the last decades, research on speciation has focused almost entirely on the development and maintenance of barriers to interspecific reproduction (i.e. hybridization), postulated to be the major driving force of speciation. Prezygotic reproductive isolation includes mechanisms contributing to ecological, behavioral and mechanical isolation between species, leading to different degrees of gene flow reduction or complete interruption. Therefore, any cues or signals capable of accurately conveying information about species, sex and reproductive status have the potential to contribute to prezygotic reproductive isolation between diverging species. Albeit studies on animal signals have been mostly devoted to visual and auditory communication, empirical evidence has accumulated that olfactory communication (i.e. chemical signaling) also plays a major role in animal communication systems. In particular, insects have exploited this type of signaling as their primary communication mode. In the following chapter, we present an overview over the contribution of chemical communication to prezygotic reproductive isolation in exemplary insect species with a particular emphasis on the variation of cuticular hydrocarbons between species and its impact on mate preference and assortative mating.

In: Animal Communication and Cognition
Editor: Tabitha Wagner
ISBN: 978-1-63482-411-8

Chapter 1

COMPUTER-ANIMATED STIMULI TO INVESTIGATE LIZARD VISUAL COMMUNICATION: A CASE STUDY IN THE JACKY DRAGON

***Kevin L. Woo*[1,*] *and Guillaume Rieucau*[2]**
[1]SUNY Empire State College - Metropolitan Center
[2]Institute of Marine Research

ABSTRACT

Visual signals that are designed for effective communication are complex. Over the years, researchers interested in the study of animal communication have employed various artificial stimuli during behavioral experiments to precisely control what the observers get to see or experience. These diverse techniques to stage animal interactions have offered an alternative to the traditional experimental methods relying on the use of live companions or demonstrators. Current technological advances allow researchers to develop realistic computer-animated stimuli of social partners with high degree of fidelity for both morphological and dynamic behavioral characteristics. Because the computer-generated animation technique offers researchers with the ability to produce standardized visual stimuli while reducing the variable

* Corresponding author: 325 Hudson Street, New York, NY 10013-1005, E-mail: Kevin.Woo@esc.edu, Phone: +1.646.230.1262, Fax: +1.646.647.7829.

behavior of demonstrators across experimental sessions, its used is becoming increasingly popular in behavioral experiments in various taxa.

In the following chapter, we provide an appraisal of the computer-generated animations efficiency to stage social interactions in lizards in general and the Jacky dragon (*Amphibolurus muricatus*) in particular. We present relevant experimental studies in which this innovative technique has been employed to simulate social partners in the Jacky dragon to address diverse ecological important questions. In addition, we consider the future of video and animation playback in the study of lizard's communication.

A Model for Visual Communication

Exciting findings over the last three decades have suggested that sensory processes form an integral part of model characteristics in signal design. Research on mechanisms hence provides insights into signal evolution. Previous studies of visual signals have focused almost exclusively on static traits, such as morphology and color; much less is known about the design of dynamic visual signals, which are defined by movement. In our research conducted over the last decade, we measured the sensitivity to movement of a particularly adequate animal model, the Jacky dragon (*Amphibolurus muricatus*), and then applied the findings to explore why and how social signals evolved particular motion characteristics. We then extended these findings by examining the way in which sensitivity to motion is also relevant for higher-order processing, such as visual signal syntax. We were able to examine signal communication in such a systematic fashion by employing computer-generated animations of artificial and realistic stimuli.

The main objectives of this chapter is to i) present an appraisal of the current knowledge on lizard communication in general, and on the Jacky dragon in particular, acquired during studies that have employed dynamic computer-generated animations and ii) identify future lines of research to further unravel the ecological and evolutionary processes that have shaped the design of visual signals in lizards.

Constraints on Signal Design

Signals that communicate efficiently between conspecifics are designed to stimulate the sense organs of intended receivers (Endler, 1992; 1993; Rowe,

1999), degrade predictably over time (Bradbury & Vehrencamp, 1998), and propagate through the environment (Ryan & Kime, 2003). Theoretical models suggest that social (Blumstein & Armitage, 1997; Leal & Rodríguez-Robles, 1997; Ord et al., 2001; 2002), perceptual (Ryan & Rand, 1993), environmental (Marchetti, 1993), or a combination of factors (Jenssen, 1977) may influence the evolution of signal design.

Constraints on signal design have influenced the evolution of its efficacy. Endler and Basolo (1998) have provided an outline on the factors of signal design during courtship, such as the sensory drive (i.e., the evolution of neural systems in response to ecological determinants: Endler, 1992; 1993), pre-existing biases (i.e., ancestral preference for particular attributes: Basolo, 1990; 1995a; 1995b), sensory exploitation (i.e., manipulation of the sensory system: Ryan & Rand, 1993), sensory traps (i.e., manipulation of neural processing: Christy, 1988; 1995), hidden preferences (i.e., training and discrimination of changes embedded in neural networks: Johnstone, 1994; 1995), receiver psychology and perceptual drive (i.e., influence of higher-order cognitive processes: Guilford & Dawkins, 1991). Two classic examples of how sensory processes have influenced evolution are the calls by male túngara frogs (*Physalaemus pustulosus*) that are used to attract females (Ryan, 1985), and the female preference for male ornamental size in the swordtail fish (*Xiphophorus* spp.: Basolo, 1990).

THE JACKY DRAGON

Jacky dragons are agamid lizards found in eastern Australia from southeast to northeast New South Wales. They inhabit visually complex environments including sclerophyll forests, coastal heathland, and rocky ridges (Harlow & Taylor, 2000). Jacky lizards are semi-arboreal, have cryptic patterning, and are diurnal. Slight sexual dimorphism is apparent. Males typically have a greater snout-vent length, are heavier, more diverse in textural pattern and color, and develop a pronounced sagittal crest at sexual maturity. They are also subject to temperature-sex determination during incubation (Warner & Shine, 2005; Warner et al., 2010).

Jacky lizards have a visual communication repertoire characterized by highly stereotyped visual displays used in social interactions, such as mate selection and territoriality (Carpenter et al., 1970; Carpenter & Fergusson, 1977). Aggressive displays are broadly similar to those of other agamids (Gibbons, 1979), and produce a display action pattern (DAP: Carpenter &

Fergusson, 1977; Jenssen, 1977) like that of *Anolis* lizards (Jenssen, 1977; Bels, 1986). Social displays comprise of three general functions: assertion, aggression, and appeasement. Assertive and aggressive motor patterns include the tail-flick (Peters & Evans, 2003a) and push-up body rock (Gibbons, 1979). Submissive motor patterns constitute a slow arm wave (Ord & Evans, 2002). They also perform a rapid backward-forward arm wave that precedes the push-up body rock, and a slow head bow also thought to be an appeasement signal (Table 1).

Table 1. Social displays of the Jacky dragon

Display Type	Motor Pattern	Visual
Aggressive	Tail-flick	
	Backward-forward arm wave	
	Push-up body rock	

Display Type	Motor Pattern	Visual
Submissive	Slow arm wave	
	Slow head bow	

We captured lizards at field sites (Table 2) located at Lane Cove, Botany Bay/La Perouse, and Royal National Parks, found within the greater Sydney metropolitan area. Consequently, we selected field locations based on known sightings of resident Jacky dragon populations.

THE DESIGN OF A COMPUTER-ANIMATED STIMULUS

Researchers have argued that video playback neglects potentially important visual information (e.g., ultraviolet colors, Bennett & Cuthill, 1994), such as the diversity in the way in which information is processed by different visual systems (D'Eath, 1998; Fleishman et al., 1998), differences in retinal physiology (Fleishman & Endler, 2000), and depth cues (Zeil, 2000). While it is generally acknowledged that video playback is tuned for the human visual system (Oliveira et al., 2000), studies have clearly shown that animals respond to video playback of conspecifics (for instance: Clark & Uetz, 1990; Evans & Marler, 1991; Rowland, 1995; Rowland et al., 1995; Kodric-Brown & Nicoletto, 1997; Rosenthal & Evans, 1998; Ord et al., 2002).

Table 2. Locations and representative photographs of Jacky lizard catching sites in three Sydney regional national parks comprised of mixed vegetation habitat

National Park	Location and Area	Map	Vegetation
Lane Cove	33°47'01"S, 151°08'19"E; 6 km^2		
La Perouse	33°98'88"S, 151°23'15"E; 4.9 km^2	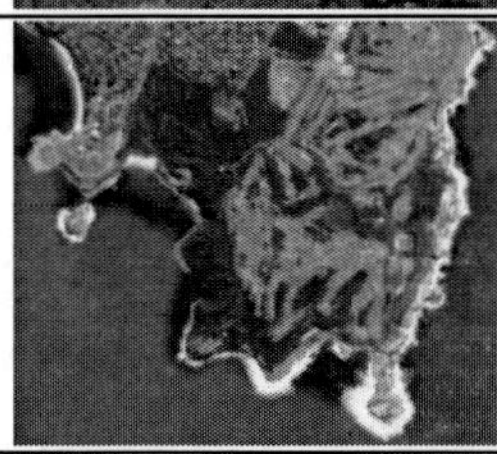	
Royal	34°7'21"S, 151°3'50"E; 154.54 km^2		

Although traditional approaches, such as using live conspecifics (Decourcy & Jenssen, 1994) or surgical manipulation of traits (Tokarz et al., 2003), has aided in understanding the evolution of signal design, the use of animations offers a more sophisticated technique for studying visual features critical for communication. In the application of this method, visual components may be isolated to better understand the sensitivity to and function of the signal (Rosenthal, 2000). Animations have been used to assess discrimination of abstract objects (Cook & Katz, 1999; Cook et al., 2001), but have also been used to examine social interactions, such as mate choice (Künzler & Baker, 1998), individual recognition (Watanabe & Troje, 2006), territoriality (McKinnon & McPhail, 1996), opponent assessment (Ord & Evans, 2002; Van Dyk & Evans, 2007), and signal design (Peters & Evans, 2003b).

As playback, if considered in any form for presenting an artificial stimulus to a real observer, has evolved from staged interactions with stationary models of varying physical features to sophisticated technology, we continue to employ new tools that enable us to ask important questions in visual communication. In our review of computer-generated animations (Woo & Rieucau, 2011), we examined the types of historically successful playback techniques, and compared each technique's limitations to each other. Although a select feature of a visual stimulus may be more salient in a specific interaction, our goal was to emphasis that computer-generated animations have the greatest range of flexibility and usability. Hence, animation models are able to accommodate more visual features and reduce the likely constraints of a signal across the different components of the visual modality.

Moreover, we further compiled a comprehensive list of studies that successfully employed computer-generated animations (Woo & Rieucau, 2011). As the first known application of computer animal models, Clark and Uetz (1990) created a crude, but lifelike animated jumping spider (*Maevia inclemems*) to see if conspecifics merely recognized the artificial stimulus, and would likely respond to it as a real one. This study provided other researchers with a potentially powerful tool in which we could create our own animations, and likely isolate specific components of the visual signal.

Video and animation techniques have been particularly successful in recent studies of communication in the Jacky dragon. Playback has been used to investigate agonistic behavior (Ord & Evans, 2002; Ord et al., 2002), discrimination based on morphology (Van Dyk & Evans, 2007), and signal design (Peters & Evans, 2003b). Moreover, in the first Jacky dragon experiments to employ computer-generated animations, Peters & Evans (2003a) used a 3D animated tail-flick to examine signal efficacy in an assertive broadcast display. In later studies, we built upon these earlier studies to employ a 3D animation of a Jacky dragon to examine signal design characteristics, and focused on conspicuousness and recognition (Woo & Rieucau, 2008; 2011; 2013; 2014).

THE JACKY DRAGON ANIMATION

To create our own realistic animation, we started with a taxidermic model of a Jacky dragon, which had died of natural causes, and used the software program Lightwave® 3D (v8, NewTek Inc., San Antonio, TX, USA). We elected to use Lightwave® 3D because of its user-friendly interface and ability to read compatible output files.

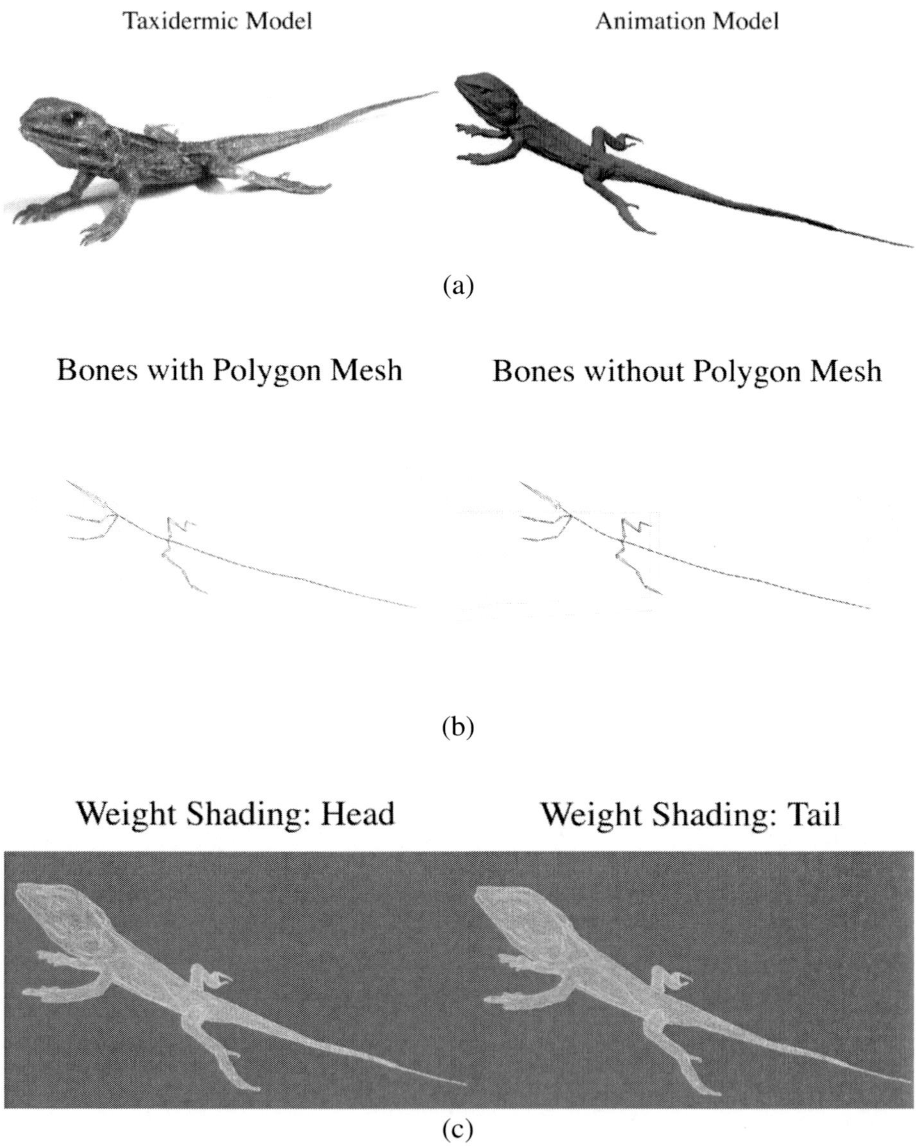

Figure 1. (Continued).

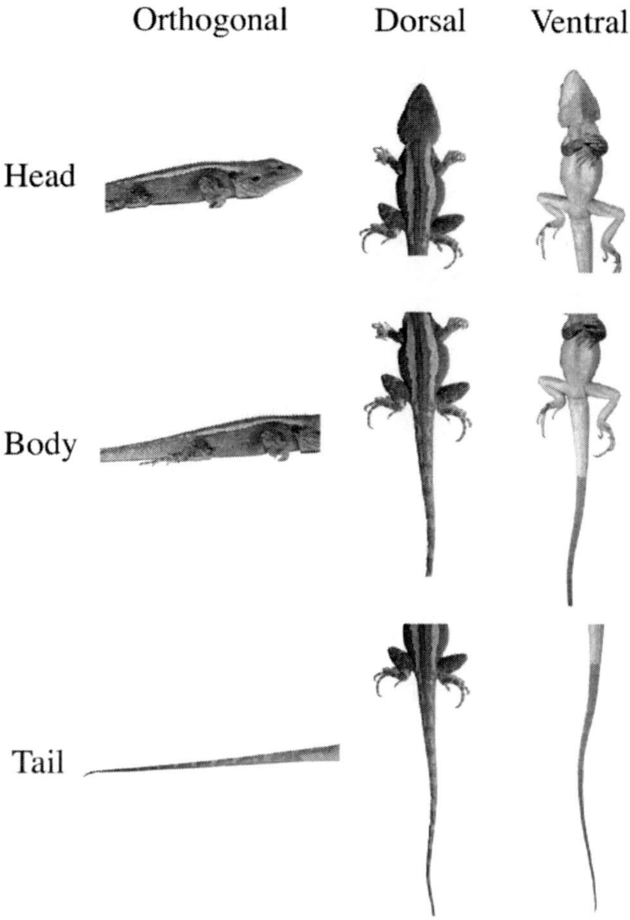

(d)

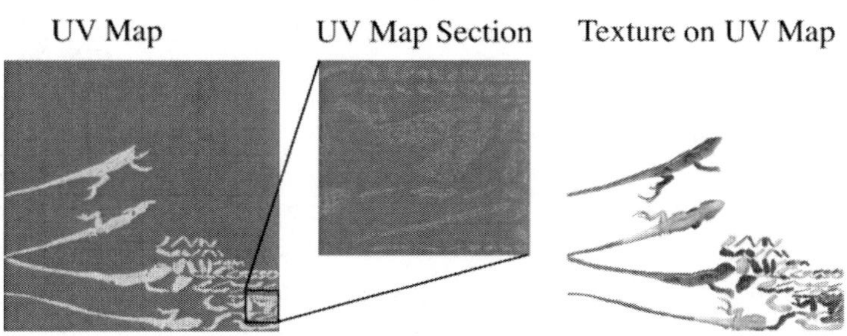

(e)

Figure 1. (Continued).

(f)

Figure 1. Stepwise aspects to create the Jacky dragon animation: a) the animation was modeled and scanned from a taxidermic specimen (left) and the animation object model (right) was created at 50,000 polygons (colored in violet); b) 61 bones were created from skelegons and the bones can be shown in the wireframe mesh to highlight the basic structure of the object without manipulating polygons; c) weight shading regions on the object by highlighting polygons - the head (left panel) was provided with a weigh shade colored in red. the tail (right panel) was given a weight shading distribution, also colored in red; d) texture was photographed from various angles and positions. We selected to show example photographs from three angles (head, body, and tail) and three positions (orthogonal, dorsal, and ventral); e) UV map to overlay the texture onto the object. We first created a UV texture map (left panel), and transferred the polygon mesh into Adobe® Photoshop® Elements, and zoomed (300%) into sections of the UV texture map to select specific pixel-to-polygon regions (middle panel); f) comparison of still video and still animation before creating movement (Woo & Rieucau, 2011 – *Behavioral Ecology & Sociobiology*).

Scanning and Lightwave® 3D

To start, we contracted a software engineering group (New Dawn®; Bexley North, NSW Australia) to scan the taxidermic, using a Konica Minolta VI-9i (Konica Minolta Holdings, Inc., Japan) lizard into a high-resolution 3D object. The Konica Minolta VI-9i uses digital photography and provides a measurement accuracy of $\pm 50\mu m$ by using a 3D algorithm to link the photographed segments together, and converts the images into 3D digital data. As a predecessor to the 3D printer, the camera photographs the model in a variety of angles, and collates the photographed images into the correct orientation and smoothing connecting contours. This technique also incorporates the use of a photogrammetry system (PSC-1) used to achieve high detail and high-accuracy of the object. Photogrammetry uses both coded markers and dimension-controlled scale bars to map the coordinates of reference markers. These coordinates constitute a 3D constellation that is used to accurate gauge the contours and distance between each photographed section. The data was collected with Raindrop Geomagic® (Raindrop Geomagic, Inc., Research Triangle Park, North Carolina USA) software, which acquires a single polygon mesh to recreate the final shape of the lizard.

In the photogrammetry output, we acquired seven variations of the 3D object comprised of 1,000 – 100,000 polygons and produced two readable formats (*.iwo and *.obj files). We chose to use the 50,000-polygon model (see Figure 1a), as this resolution provided us the ability to achieve the highest level of detail and the trade-off between computational capabilities. More importantly, the output files were compatible with Lightwave® 3D.

Lightwave® 3D comprises of two separate program interfaces that collectively complete an animation sequence: Modeller and Layout. Lightwave® 3D Modeller allows for the manipulation of the object by highlighting specific polygons for changes, and creates layers to the object, adds color and texture, and creates skelegons. Lightwave® 3D Layout creates scenes that incorporate all the independent objects built from the Modeller. In both Modeller and Layout, the object has three dimensions of orientation, positioning, and rotation (see Figure 2): pitch (X), heading (Y), and bank (Z).

Skelegons and Bones

In Lightwave® 3D Modeller, we were able to imbed 'skelegons' into the object. 'Skelegons' act as placeholders for virtual bones in Lightwave® 3D Layout and create a hierarchy for movement (see Figure 1b). For our model, we built 61 skelegons. For ease of inserting skelegons, we opened another layer in the program to view the object as a wireframe. In Modeller, we can view multiple wireframe layers that prevent accidental highlighting or movement of polygons while we created skelegons. We created an artificial spinal column to simulate the cervical vertebrae from the neck down to the sacral vertebrae to the tip of the tail, thereby recreated the skeleton of a Jacky dragon. We used one large skelegon to simulate the head. Each forelimb consisted of fours skelegons each and we then fused them to the spinal column. The skelegons were fused together creating a hierarchical system where the spinal column acted the central foundation for all limb movements. After we inserted all skelegons, we synchronized the object to Lightwave® 3D Layout and converted the skelegons to bones.

Weight Maps (Vertex Mapping)

In Lightwave® 3D, weight maps have a general value range from –100% to +100% distribution of motion. For instance, independent weight maps designated to specific areas act antagonistically to allow smooth and realistic movement of the object. Weight values that deviate from 0% (no effect) will produce a greater effect of movement on the body part and the movement of bones. The failure to weight properly may inhibit or restrain the movement of the object in relation to bone movement (i.e., bones may protrude from object with object movement in the same general direction) or hyper-movement (i.e., movement of object may supersede the position of the bones in a general direction). In our model, limbs, tail, and head were designated with positive (+) weight map values, while the body was designated a negative (-) weight map value (see Figure 1c). We maintained weighting at relatively low values $\leq \pm 25\%$ for all antagonistic parts to prevent exaggerated movement.

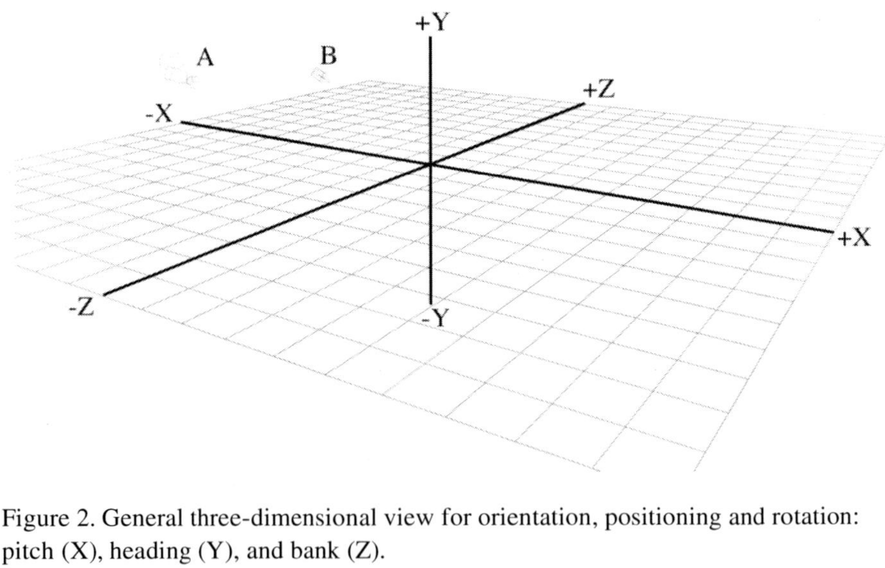

Figure 2. General three-dimensional view for orientation, positioning and rotation: pitch (X), heading (Y), and bank (Z).

Texture

To acquire the perceived texture of the lizard, we photographed the patterns of a live Jacky dragon similar to the taxidermic model in its mass and length. We photographed a live lizard was photographed with a Canon EOS 5D (Canon Inc., Toyo, Japan) from two angles (i.e., frontal and orthogonal), from four positions (i.e., frontal, orthogonal, ventral, and dorsal), and from five sections (i.e., whole animal, head, body, tail, and limbs) over a white sheet of paper and balanced for pure white RBG values (see Figure 1d).

First, we imported the polygon mesh into Lightwave® 3D running on a Power Mac G5 (Apple Computer Inc.). To place skin texture on the model, we then created a UV texture map (see Figure 1e). This process assigns two extra coordinates to every point on the model that are represented as the two axes of a plane. In effect, this peels the surface off the model and spreads it flat as a 2D image. The texture is placed on a particular point on the UV map, and it is automatically transferred to the corresponding point on the model. We then imported each section of the skin texture into the UV texture map using Adobe® Photoshop® Elements 3.0 (Adobe Systems Incorporated). The match between the UV map and sections of skin texture was not always exact, and occasionally left a small fraction of the points uncovered. When this occurred, we placed the color from the nearest point on the uncovered point using the 'eyedropper' and 'paintbrush' tool.

Image Sequence Capture

Prior to the creation of our animation, we captured simulated male-male interactions as video sequences. To do so, we used a Canon MV650i digital camcorder (Cannon Inc., Japan). We placed male Jacky dragons in adjacent and isolated enclosures (~1m apart) and we filmed them for social displays. We strategically placed wooden perches, with a heat lamp above the perch, at an orthogonal position to the camera to draw lizards onto the perch. We matched the width of the perch exactly to the size it would be represented on the video monitor.

We selected motor patterns (e.g., tail-flick, push-up body rock, and slow arm wave) from the captured digital video footage and exported the segmented sequences into an image sequence (series of consecutive JPEG files) using Apple QuickTime™ Player 7.0.4 (Apple Computers Inc.). We exported the video sequences as PAL-DV standard (720 x 576 pixel resolution at 25 frames

s^{-1}), which represents the regional coding for Australasian countries. PAL-DV standard must be maintained across each step (i.e., sequence capture, rotoscoping, and rendering) for frame rate continuity.

Rotoscoping

As an effective tool for matching movement patterns, 'rotoscoping' is a technique where the model is superimposed onto a background image or series of images (Gatesy et al., 1999). Here, we transferred the object (see Figure 3) from Lightwave® 3D Modeller to Layout, and retained all design features. In Layout, we controlled the environment, and established parameters for the camera (e.g., field of view), light, object, and background characteristics.

We imported the first JPEG image into the background of the camera view. We manipulated the model using the motion parameters of the bones, and superimposed on the background image (see Figure 3). The position is then 'keyframed', which saves the orientation of the bones. Following the first keyframe, we replace the JPEG in the background with the next consecutive picture in the image sequence and repeated the process. Lightwave® 3D Layout used a motion algorithm to calculate movement pathways in between keyframed positions. When the scene is complete, the sequence was exported into an image sequence and rendered to create one sequence.

Rendering

To a large quantity of individual JPEGs prior to conversion into a movie, we needed to render the images from Layout. We used Render Farm Commander v2.9 (Bruce Rayne©, Lalor Park, NSW Australia) to render all large sequence clips. Render Farm Commander (RFC) employs all computers connected on a local area network system to increase rendering time by distributing jobs among linked computers. We used four Apple Mac G5 dual-processors (eight threads) render our sequences. For example, we processed 9000 frames (i.e., equivalent to six minutes on PAL-DV standard) in 12 hours using a single G5 processor and reduced the time to four hours when redistributed across eight threads (i.e., four G5 dual processors). The use of RFC for batch processing is efficient when there are more than two large sequences. We chose to render both long and short sequences into individual JPEGs.

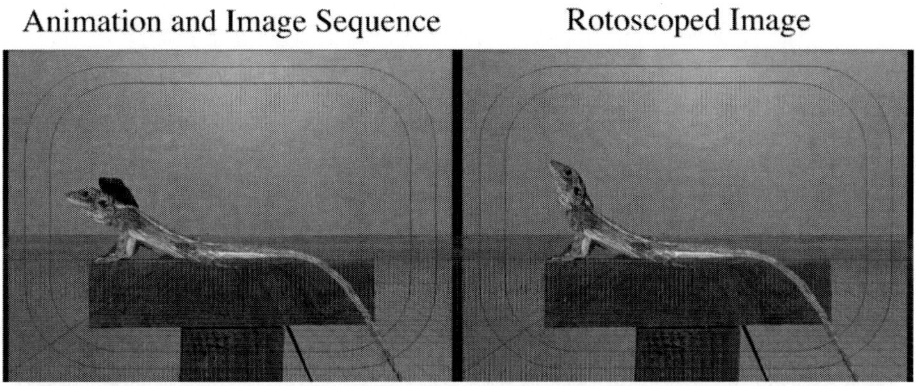

(a)

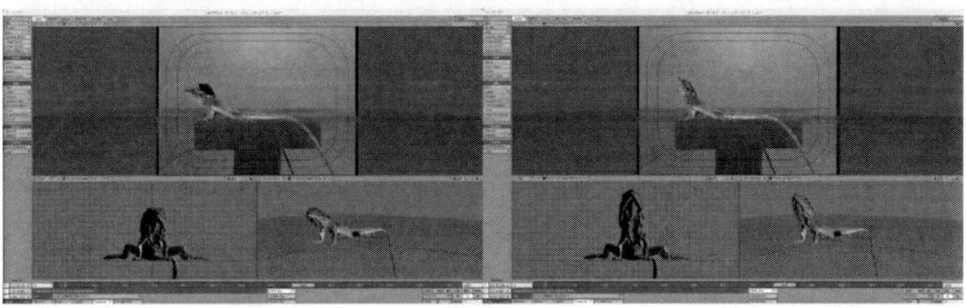

(b)

Figure 3. Rotoscoping the animation in Lightwave® 3D from image sequences exported as individual JPEGs. a) The animated lizard is superimposed in front of the first JPEG in the image sequence of a push-up body rock (left panel). The animation is then matched to the same position by manipulating the bones imbedded within the object (right panel). b) The bones are seen in the 'X-ray Bone' view that allows us to see the skeleton within the object (colored in blue). The background image is changed to the second JPEG in the image sequence, and the object is manipulated in to the new position. This processes is repeated until the sequence is completed. (Woo & Rieucau, 2011 - *Behavioral Ecology & Sociobiology*).

Making the Movie

Creating movies for playback can be done in two ways using: 1) Final Cut Pro (Apple Computers Inc.) and 2) Apple QuickTime™ Player. With large sequences, we exported each JPEG image sequence into Apple QuickTime™ Player 7.0.4 and converted the sequence into a DV stream. Lastly, we imported each individual DV stream into Final Cut Pro to create a movie sequence.

Optic Flow Analysis on Stimulus Design

Optic flow analysis measures the change in speed and velocity of an object by detecting the changes in image contrast over time. Motion detection is dependent on the ability of the visual system to adjust to changes in velocity from the apparent movement between proximate and distal cues (Zeil & Zanker, 1997). For example, as honeybees (*Apis spp.*) navigate through novel environments (i.e., artificial mazes), they have the ability to calculate their own motion pathways in comparison to the relative movement of objects (Srinivasan et al., 1996).

To ensure that our models were lifelike in both their design and motion patterns, we established a method to compare the video sequences to their animation counterparts (Woo & Rieucau, 2008). It was important to quantitatively measure our animations to ensure that conspecifics that responded naturally to video may also respond similarly to our computer-generated sequences. We employed the matrix laboratory based (MATLAB) program, Analysis of Image Motion, (AIM) created by Richard Peters (Peters et al., 2002). AIM uses an optic flow algorithm to measure the change in motion patterns by examining the change in speed and velocity in an active space over time. We selected our representative video sequences and corresponding animation sequences, and processed them through AIM. Here, the program produces standard ellipses and velocity signatures of each clip.

Jacky lizards will broadcast all displays in any orientation. To ensure that a signal is likely conspicuous, noticeable, and likely recognized as a specific signal, we compared the videos of a single display (e.g., slow arm wave) in two different orientations: frontal and orthogonal. Consequently, our aim was to find no significant difference between displays as evidence that our animations were as realistic as the video. We then compared the video and animation sequences for three social displays (tail-flick, push-up body rock, and slow arm wave) for their likely consistency (Table 3), and likewise found no significant differences between them (Figure 4). As it seemed likely that both the video and animation sequences produced the same standard ellipses and velocity signatures over time, this suggested that we could now stage interactions using our animations, and real conspecifics would indeed respond to them.

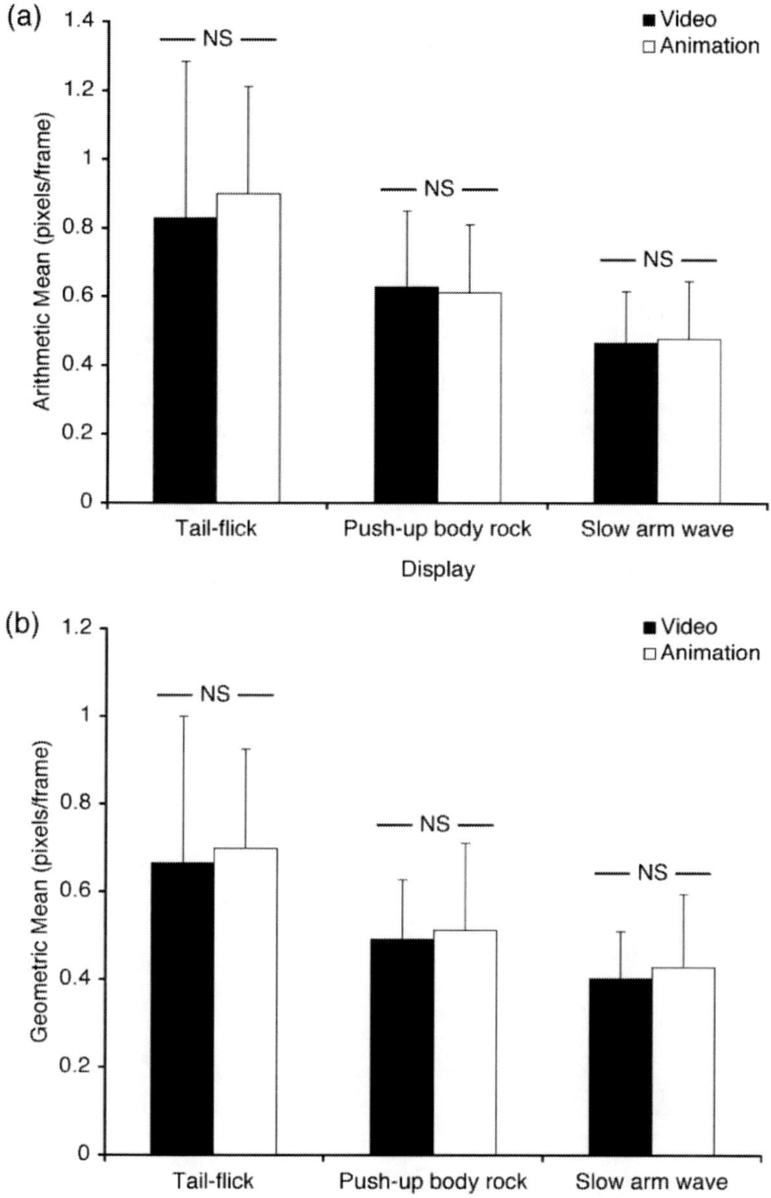

Figure 4. In comparing the video and animation sequences in Analysis of Image Motion, there was no significant difference between the average change of pixels per frame, which suggested that the animations were created very similarly to their video counterparts (Woo & Rieucau, 2008 - *Behavioural Processes*).

Table 3. Comparative standard ellipses and velocity signatures, as generated by Analysis of Image Motion (AIM), which showed no significant differences between video and animated sequences (Woo & Rieucau, 2008 - Behavioural Processes)

		Signal Design Optic Flow Output	
		Standard Ellipse	Velocity Signature
Tail-flick		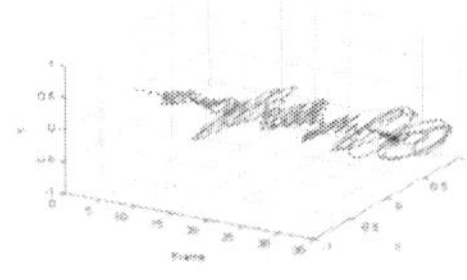	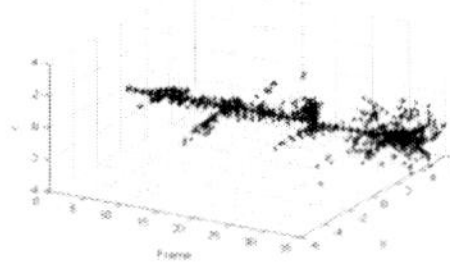
	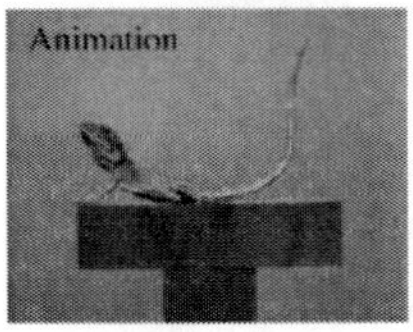	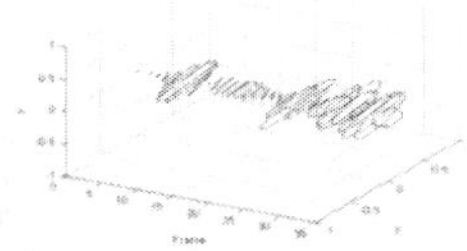	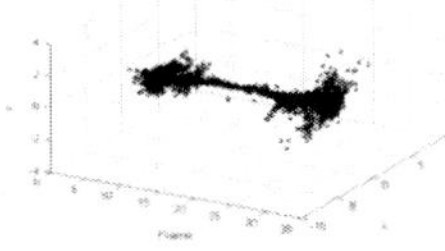
Push-up body rock		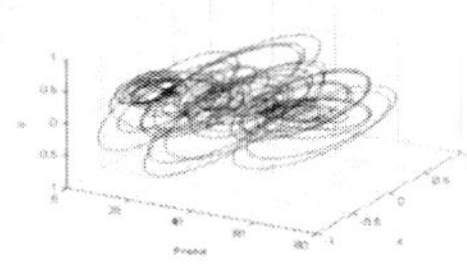	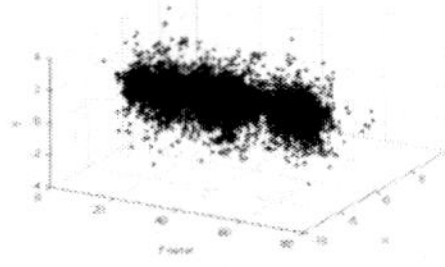
		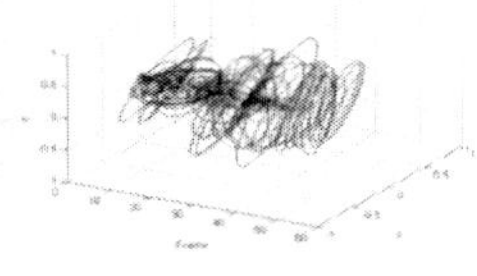	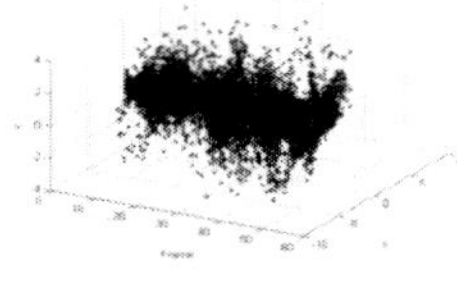
Slow arm wave		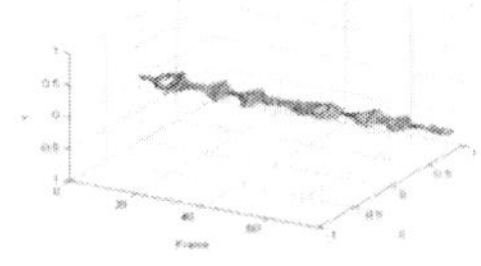	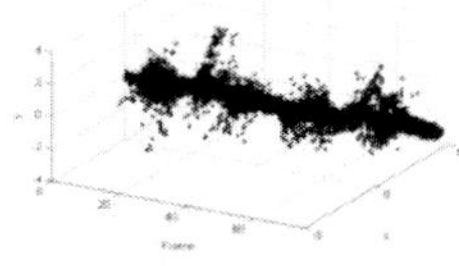
		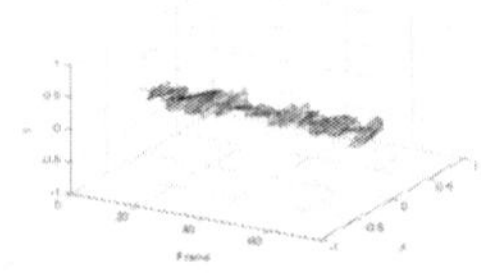	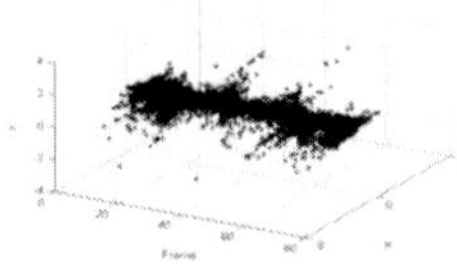

EXAMINING SIGNAL CONSPICUOUSNESS

Sensitivity to motion is a critical feature in visual discrimination (Clifford & Ibbotson, 2003). In order to deconstruct the parameters of motion within each display, we first examined the basic perception to movement. By measuring the visual threshold in its range of speed and coherence, we could then match motion patterns that are likely perceived by the Jacky dragon visual system. Sensitivity to stimuli that are inherent of defined motion correlates to the relatively important signals within the display repertoire. Consequently, it also elucidates the types of motion that also irrelevant, and likely outside the perceptual threshold. Thus, to determine the motion threshold of Jacky dragons is the first step to understand the overall importance of all displays.

To measure relative sensitivity, we examined conspicuousness with two complimentary kinds of studies. First, we employed computer-generated random-dot kinematograms to measure basic sensitivity to motion. Secondly, we coupled our data with experiments that used the Jacky dragon animations, which were designed with the motion characteristics within observable patterns of movement.

Motion Sensitivity and Random-Dot Kinematograms

Classically, motion sensitivity has been examined through the establishment of critical flicker fusion (CFF) rates, which is the maximum frequency at which particles of light can be perceived as discriminate (Lythgoe, 1979; Davson, 1990). Humans have a CFF rate of 50-60 Hz (Sekular & Blake, 1994). In comparison, starlings (*Sturnus vulgaris*: Maddocks et al., 2001) have a high CFF of >100 Hz, while cane toads (*Bufo marinus*: Nowak & Green, 1983) have a low CFF rate of ~6.7 Hz (see Woo et al., 2009b for representation from comparative species).

Sensitivity to moving objects in an organism's environment is paralleled to the mechanism that exploits the visual processing system. Understanding this mechanism provides a basis in which to examine the sensory capacity involved in signal evolution of a species. Signal evolution suggests that signals have been designed to propagate through the environment, stimulate receiver sense organs, and carry information important to both the receiver and signaler (Bradbury & Vehrencamp, 1998). The movement-based signals of the Jacky dragon (*A. muricatus*) have been designed to meet the constraints of signal design within a complex environment through the evolution signal-receiver systems of perceptual, environmental, and social factors (Ord et al., 2002). Rapid and distinct movement characteristics such as direction, speed, acceleration and sweep area serve as an alerting mechanism to conspecifics (Peters & Evans, 2003b). Sensitivity to these motion vectors have been critical to male-male interactions when competing for territory or potential mate (Carpenter et al., 1970), and failure to perceive these signals may result in the inability to compete for resources. However, this sensitivity to social communicative displays may have also neglected cues that exploit other sensitivity ranges.

In order to test signal speed and strength, we used random-dot kinematograms (RDKs). RDKs are a classic psychophysical tool used to measure motion thresholds in the visual system, highly salient in engaging the visual system, do not engage recognition mechanisms, and do not represent any biologically relevant stimulus (Curran & Braddick, 2000). Burghardt (1977) suggested that instrumental learning paradigms are used as tools to test learning phenomena as well as sensory processes. By using psychophysical procedures, we test perceptual learning and RDKs to gauge motion sensitivity characteristics such as orientation and discrimination (Matthews et al., 1999), direction (Liu & Weinshall, 2000), and coherence, direction, luminescence, (Zanker, 1999). In understanding the nature of coherent motion, Kingstone and Bischof (1999) have proposed three general rules: 1) coherence discrimination is easier with fewer motion groups, 2) visual inhibitory mechanisms allows for more discrete discrimination over excitatory mechanisms, and 3) the greater number of coherent objects, the more coherent the motion has become. These rules underline the use of coherence as a mechanism for perceiving the relative strength of the signal. Signal strength coupled with speed vectors either increase or degrade depending on an individual's ability to correctly discriminate. By using psychophysics as a technique to measure sensory perception, it incurs an acquisition period in which the subject has to learn how to perform a certain task. Historically, shaping herpetological species for an instrumental procedure have proved difficult due natural mechanisms attributed to physiology, responsiveness, performance ability, and the type of reinforcer that is used (Burghardt, 1977). Psychophysical measures require subjects to discriminate between two or more perceptually different stimuli.

We used RDKs as a neutral stimulus in which to signal secondary reinforcer. Movement, shape, and color of the dots do not represent any biologically inherent recognition, and therefore the selection to use dots was justified because it did not trigger any salient responses that would be elicited at the presentation of stimulus that was inherently biological. By implementing a basic instrumental procedure, we can gauge the level of motion sensitivity according to the responses made in viewing the artificial stimuli. They have now been used in a series of successful psychophysical studies of animal and human visual perception (Newsome & Paré, 1988). We measured motion sensitivity by generating variations in RDK speed (degrees per second) and coherence (percentage of RDKS moving in the same direction; Figure 5). The capacity to resolve moving dots may provide insight into the operation of the visual system, and its sensitivity to speed (Vreven & Verghese, 2002) and coherence (Verghese & McKee, 2004) levels. Discrimination of the RDK features may also help to identify how the visual system separates the motion patterns of ecologically relevant stimuli (Bischof et al., 1999) from irrelevant complex visual noise (Scase et al., 1996).

Although most motion coherence studies examine human perception, the use of RDKs to examine sensory perception is applicable to non-human species (Lazareva et al., 2005). This technique has been used across a small range of non-human taxa such as pigeons (*Columbia livia*: Goto et al., 2002; Ushitani et al., 2004), macaques (*Macaca nemensrina*:; Kiorpes & Movshon, 2004), domestic cats (Burnat et al., 2002), and barn owl (*Tyto alba*: Willigen et al., 2003). Responses to these psychophysical measurements mirror our understanding of how species perceive biologically relevant motion as a function of signal evolution.

We measured the response to the interaction between speed and coherence in order to gauge the motion sensitivity of the Jacky dragon to moving stimuli. Initially, we produced independent measures of examining the interaction between speed and contrast. In addition, the use of archival displaying lizard footage proved as an unsuccessful primary reinforcer due the unknown social implications provided by the lizard signals. To reduce the elicitation of negative responsive behavior by the subjects (i.e., hyperactivity or submissive displays), we implemented a basic instrumental procedure in which to test motion sensitivity using the parameters of speed and directional coherence (signal strength). This psychophysical procedure allowed us to gauge the level of visual perception in which the Jacky dragon is able to detect motion by measuring the responses to directional coherent motion at various speeds (Figure 6). By understanding the way motion is perceived, we can link the mechanisms associated with motion detection to biologically inherent motion vectors in the Jacky dragon visual system. The greater implications of interactions between motion vectors does not only identify visual cues in which an organism can choose to ignore, but salient motion patterns that are perceived as important such as the design of communication signals. These implications provide the basis for our understanding of signal evolution in the species.

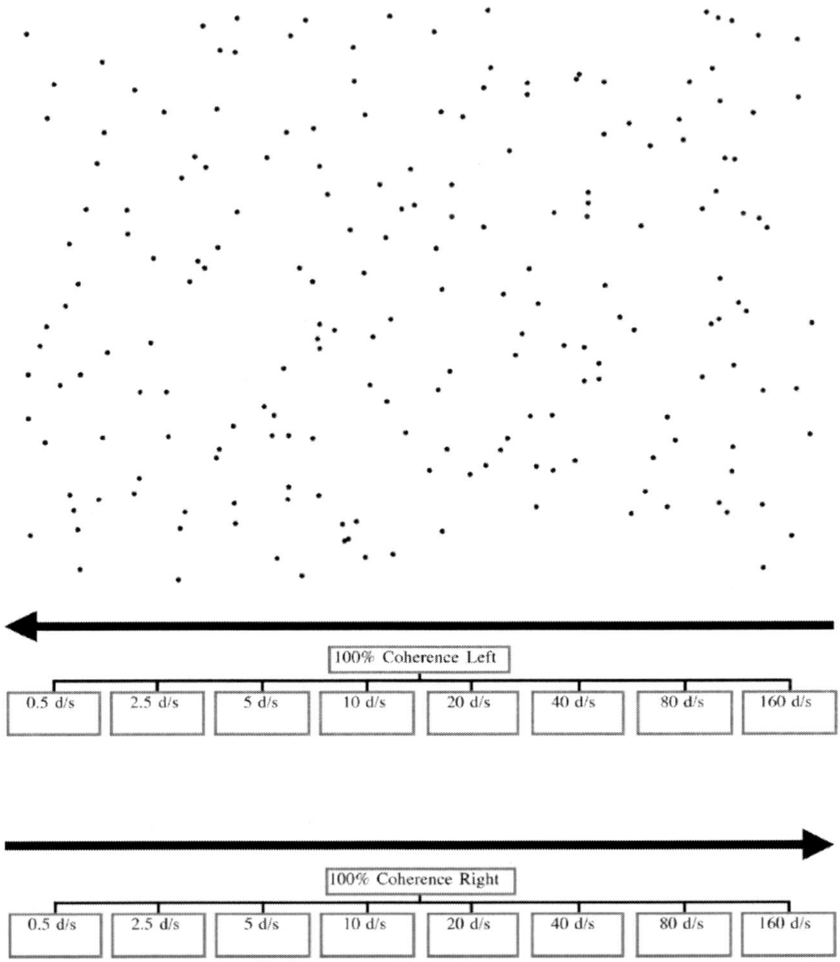

Figure 5. Representative RDK stimuli and the likely number of combinations in three factors: Coherence (0%, 2.5%, 5%, 10%, 20%, 40%, 80%, 100%), Speed (0.5 d/s, 2.5 d/s, 5 d/s, 10 d/s, 20 d/s, 40 d/s, 80 d/s, 160d/s), and Direction (left, right).

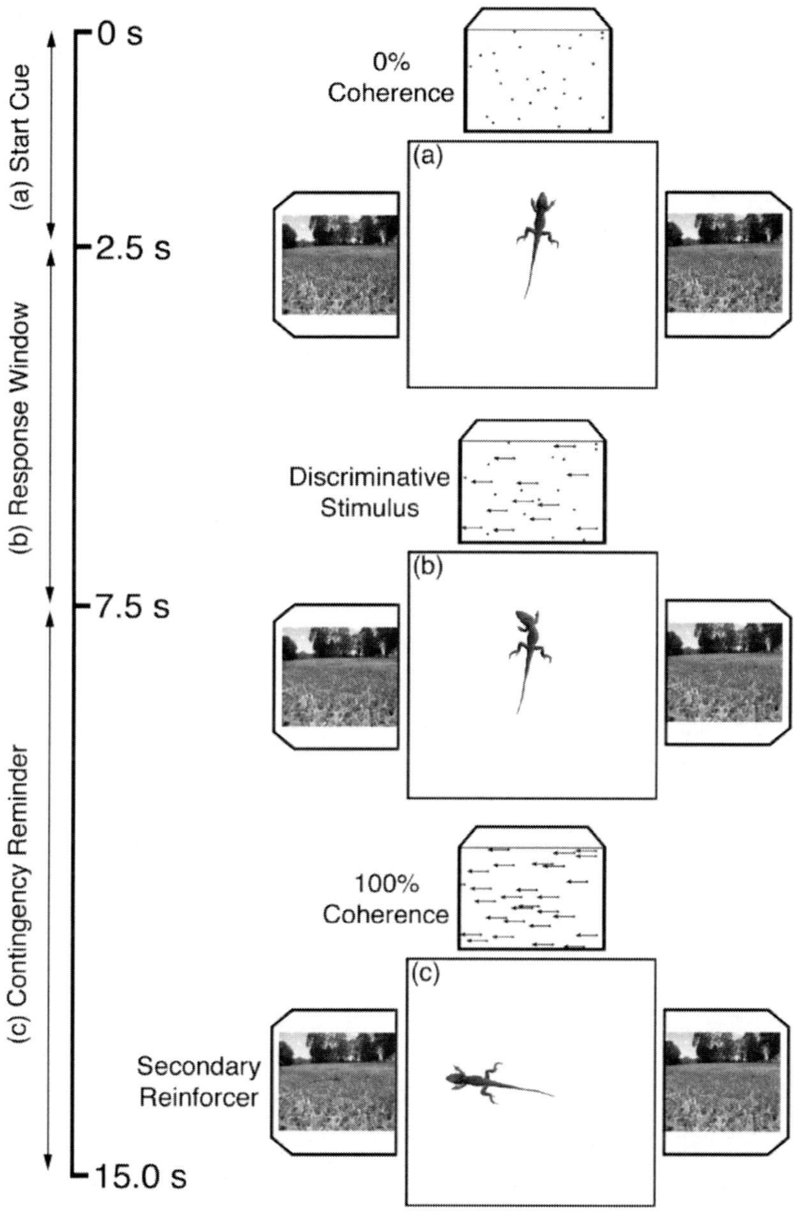

Figure 6. Timeline for test trial showing subject orientation to the correct monitor using RDK motion characteristics.

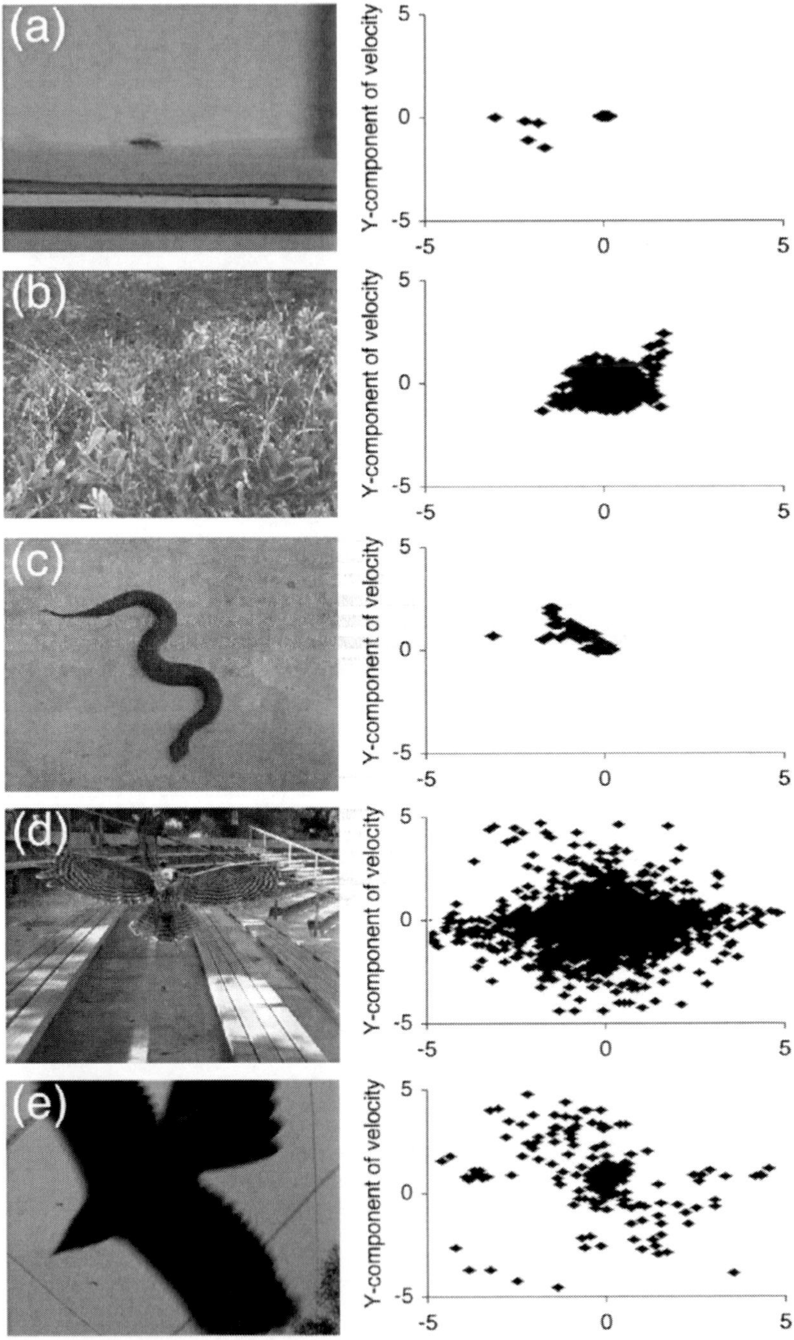

Figure 7. (Continued).

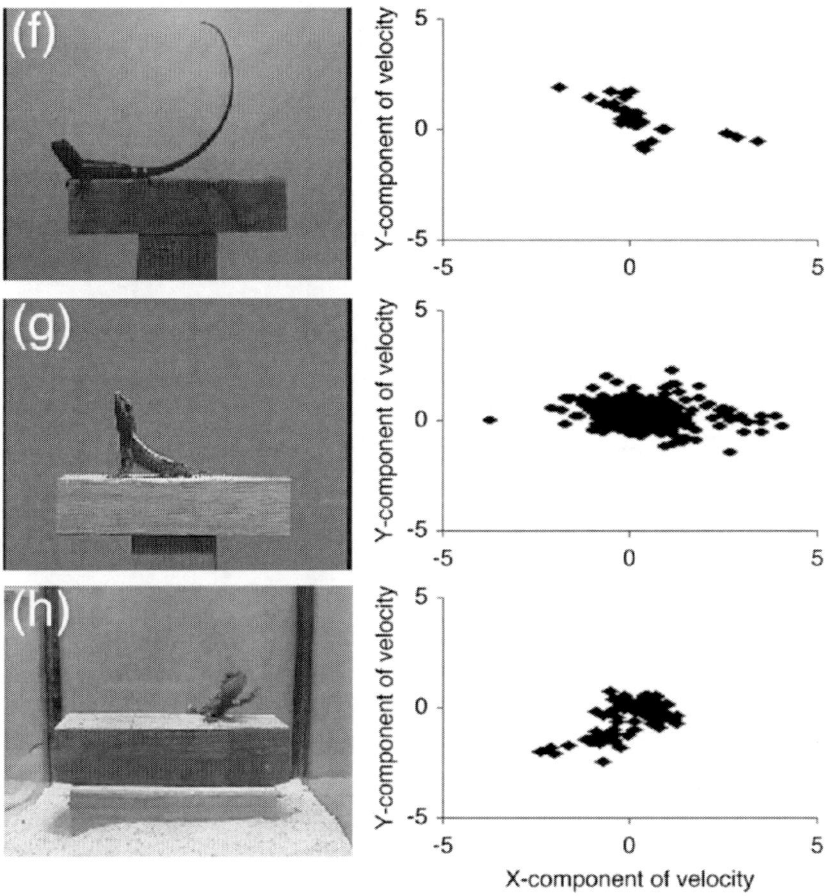

Figure 7. Generation of speed profiles using AIM: a) cricket (N = 5); b) background vegetation (N = 10); c) snake (N = 10); d) aerial approach (N = 16); e) aerial overhead (N = 36); f) tail-flick (N=6); g) push-up body rock (N = 6); and h) slow arm wave (N = 6). Generation of speed profiles produces two segments: video image (left), and velocity signature (right).

We measured the motion vectors of ecologically important stimuli using an optic flow algorithm to measure speed and velocity profiles based on the average change in pixel movement within a spatial field (AIM; Peters et al., 2002). In addition to the social displays, we included prey, avian predators, ground predators, and environmental noise, such as windblown vegetation (Figure 7). Speed sensitivity to RDK motion was then matched to the properties of ecologically relevant stimuli, focusing particularly on signals used for social communication, as well as prey escape and predator behavior.

Peters et al., (2008) also analyzed as an agent that masks movement, and as potential factor that has shaped signal design.

As speed increased, so did the Jacky dragons' ability to perceive constant motion in a coherent direction (Figure 8). This coupling effecting suggested that higher speeds and coherent motion were important for signal detection, whereas slower speeds and random direction was difficult to resolve a pattern. We then matched both speed and coherence to the ecologically relevant motion patterns generated from our AIM data (Figure 9). Jacky dragons are sensitive to high-speed motion, such as those produced as social signals by conspecifics and the movement patterns of aerial predators. Despite this ability to resolve high-speed characteristics, lizards have substantial difficulty in distinguishing slow speed movement, such as that of windblown vegetation. The evolution of high-speed movement used to stimulate the receiver visual systems suggests that sensory exploitation is likely a constraint on signal design. The results are consistent with the possibility that high-speed display movement exploits pre-existing sensitivity selected for the need to detect aerial predators. Consequently, a prediction can be drawn that suggests prey and predators potentially use the inability of Jacky lizards to resolve slow speeds to their advantage. For prey species avoiding capture and ambush predators that want to advance closer to lizards, adopting movement characteristics similar to that of windblown vegetation may greatly reduce the probability of detection.

Influence of Visual Noise

To measure visual sensitivity to salient movement, psychophysical tools (i.e., RDKs) and motion algorithms (i.e., AIM) can assist in identifying constraints on motion perception (Endler, 1987, 1992; Fleishman, 1988a, 1988b; Ryan & Rand, 1993). In the first RDK test, we were able to identify the motion thresholds for visual sensitivity. However, the ecosystem envelope for the range of Jacky dragons includes a range of environmental noise that likely conflicts with the transmission of signals.

Environmental noise influences the design of communication signals (Morton, 1975; Wiley & Richard, 1978; Ryan & Brenowitz, 1985; Ryan et al., 1990). The visual system must discriminate important signals from irrelevant ecological noise, such as windblown vegetation (Fleishman, 1992). In lizards, DAPs are stereotyped signal sequences produced during interactions with conspecifics engaged in territorial disputes (Bels, 1986).

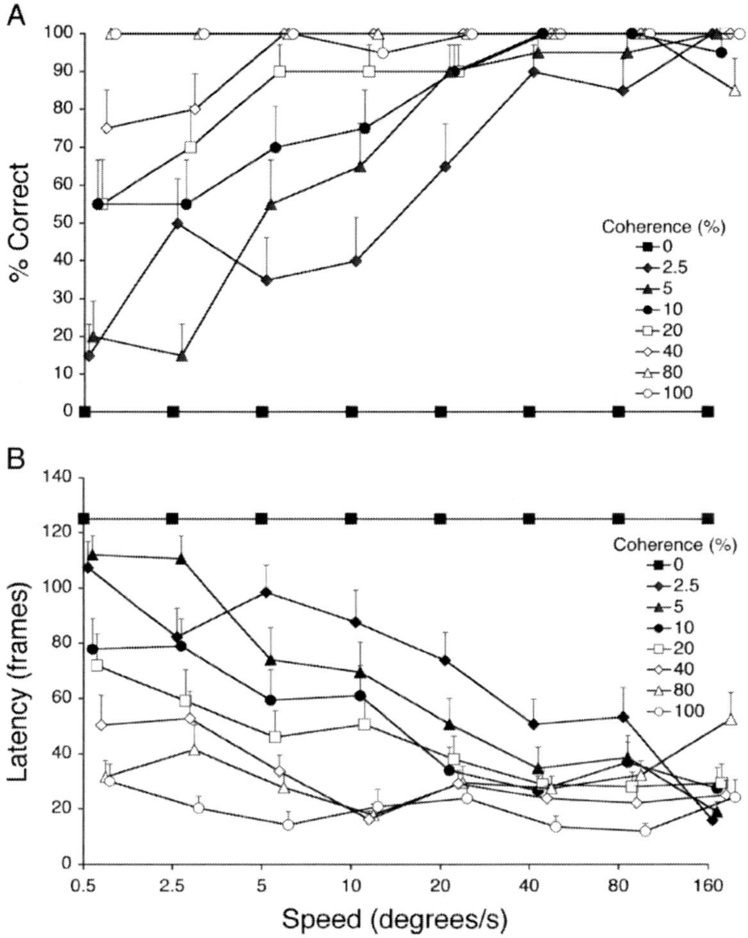

Figure 8. Performance of subjects across coherence levels (0%, 2.5%, 5%, 10%, 20%, 40%, 80%, 100%) as a function of speed, showing a main effect of both speed and coherence, as well as an interaction between the two factors: a) increase in mean percent correct responses (mean+SE) with speed; b) decrease in latency of responses (mean+SE) before reaching asymptotic performance at 40°/s). Coherence levels are identified as 0% (■), 2.5% (◆), 5% (▲), 10% (●), 20% (□), 40% (◇), 80% (△), and 100% (○). Data points were staggered for clarity.

Fleishman (1988a; 1988b; 1992) suggested that the *Anolis* visual system filters irrelevant complex noise generated by the movement of vegetation in wind and separates important visual cues, such as DAPs from neighboring conspecifics, which are more prominent than the masking effect of environmental noise. Thus, for signals to be conspicuous against visual noise, individual must be able to monitor changes in microhabitat induced by windblown vegetation. In some cases, animals that inhabit complex environment (Jenssen & Swenson, 1974; Fleishman & Persons, 2001) have also evolved a design component to increase signal detection and to prevent a masking effect by windblown vegetation (Peters, 2008).

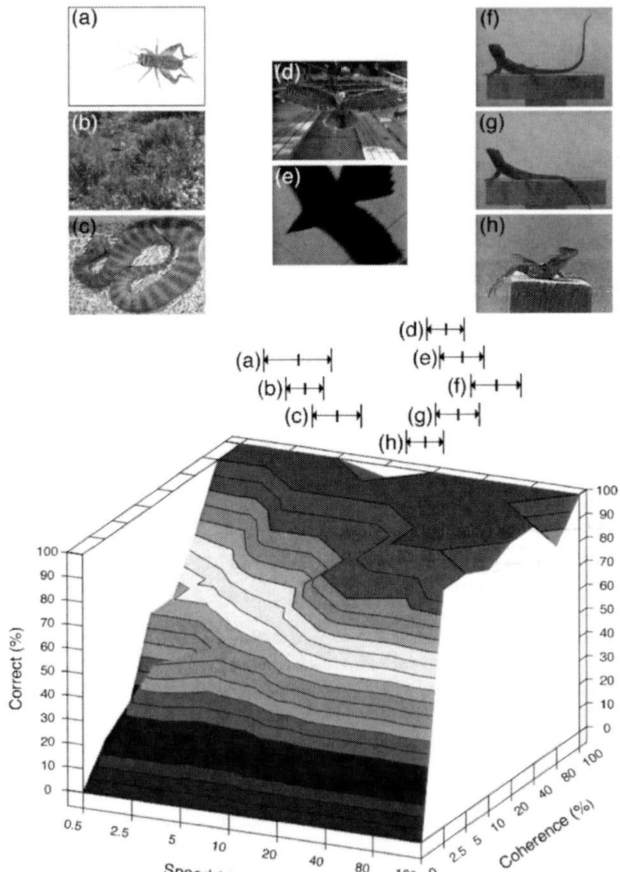

Figure 9. Summary of motion sensitivity results, despite the interaction between speed and coherence. Optic flow analysis revealed that the low sensitivity (slow movement) region included: (a) invertebrates, (b) snakes, and (c) vegetation movement. High sensitivity (fast movement) region included (d) approaching and (e) overhead aerial predators, together with all social displays: (f) tail-flick, (g) push-up body rock, and (h) slow arm wave.

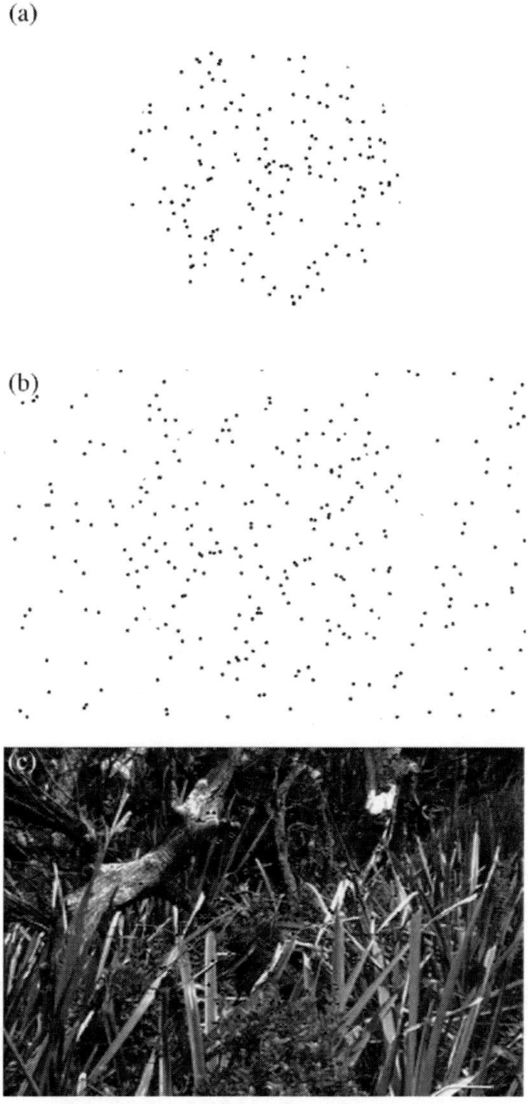

Figure 10. Examples of RDKs superimposed over three simulated backgrounds: a) none (white); b) artificial; and c) natural (Woo, Burke, & Peters, 2009 – *Animal Behaviour*).

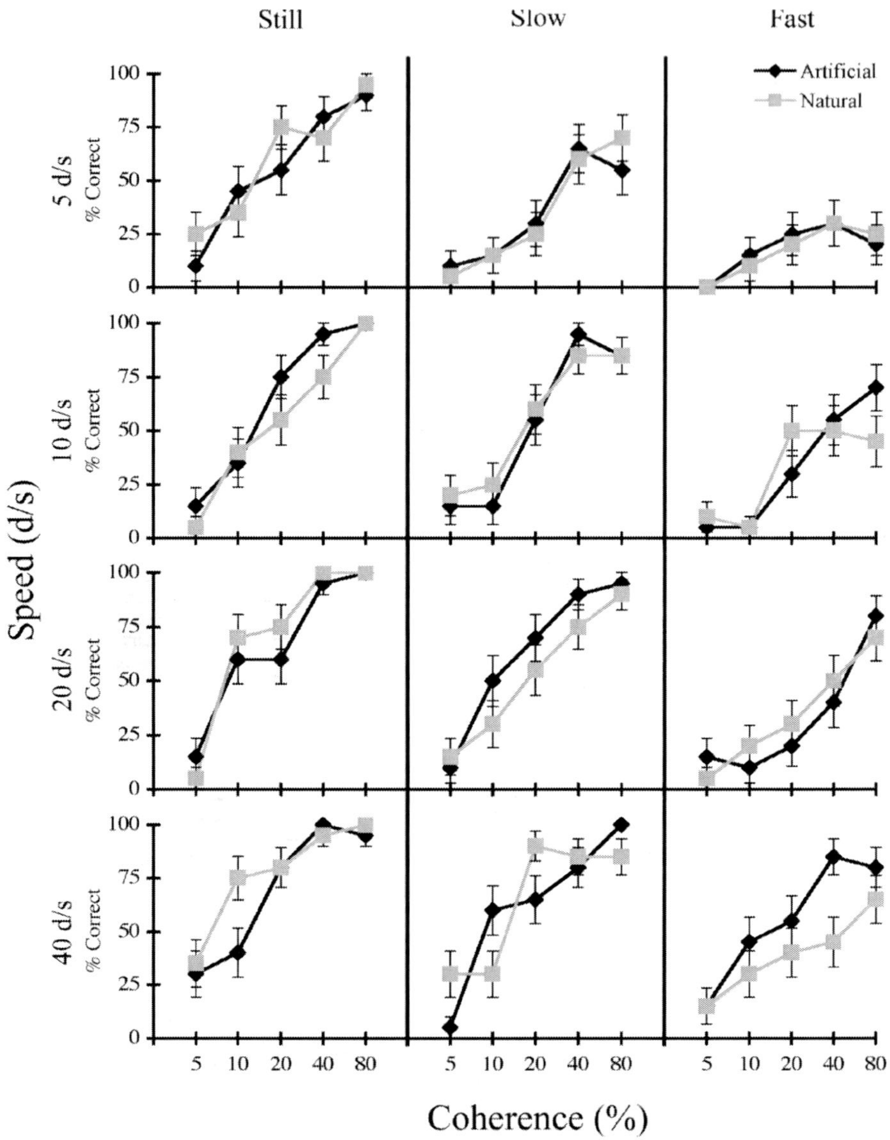

Figure 11. Accuracy of response showing an interaction between speed and coherence that is consistent across wind conditions (Woo, Burke, & Peters, 2009 – *Animal Behaviour*).

To then account for the effect of environmental noise, we embedded the RDKs within the speed and coherence range of social displays within windblown vegetation (Woo et al., 2009a). This study considered a more realistic aspect in which to understand how signals are constrained by environmental noise, as it is highly unlikely that signals are produced in the absence of other motion patterns in the background or foreground. Hence, the superimposed RDKs on top of video footage of windblown vegetation simulated real interference between important and irrelevant motion (Figure 10). Peters et al. (2008) measured the motion of different kinds of background vegetation, and we used the values to simulate the movement of background vegetation within a normal range.

The interaction between the social displays speed and coherence, and the speed and coherence values of background vegetation showed an important relationship in signal detection. Increased vegetation movement clearly compromised the ability to perceive relevant motion (Figure 11), However, as the upper range of normal windblown vegetation peaked, and the speed and coherence of RDKs progressed above the threshold into the realm of normal display motion, lizards were clearly able to detect the dots. The results showed further support for the design of signals within an optimal motion range.

Figure 12. Image of animated jacky dragon perched at a perceptual 3m in depth (Woo & Rieucau, 2012 – *Ethology*; Woo & Rieucau, 2013 – *Ethology, Ecology & Evolution*).

Signal Design and Employment of the Jacky Dragon Animation

Prior to our RDK studies, Peters & Evans (2003a) employed the first lizard animation to examine the conspicuousness of the tail-flick as an alerting signal. In their study, Peters & Evans (2003a) created an animated tail only, rotoscoped its movement based on archival video, and superimposed the tail in front of general footage of windblown vegetation. Set at a perceptual distance, they measured the conspicuousness of the tail-flick in the orientation responses of the intended receiver. This was the first successful demonstration of the use of the lizard computer-generated animation.

Following the design of Peters & Evans' (2003a) study, we also measured the conspicuousness of the push-up body rock in comparison to the tail-flick (Woo & Rieucau, 2012). Unlike the study by Peters & Evans' (2003a), which only used the tail, we used the entire Jacky dragon animation, and isolated the desired motion patterns of interest. However, we also maintained the importance of environmental noise, and subsequently superimposed the animation in front of windblown vegetation. To simulate signal production as a broadcast display, we situated the animation at a perceptual 3m in distance (Figure 12). Furthermore, to allow the focal lizard time to desensitize to the onset of video, we structured the timing of each stimulus. (Figure 13). In our study on push-up body rocks, we examined the likely constraints on the number of consecutive push-up body rocks, which may be produced individually or in bouts of five consecutive displays. Similar to our design for the basic animation sequence, we reviewed archival footage, and measured the duration of displays, frequency of individual push-up body rocks, and how many were produced within each bout. We then designed our exemplars to produce one or five displays. Our results produced a clear match for the optimal number of individual displays to the highest likelihood of its conspicuousness. Moreover, the orientation response to the mean number of push-up body rocks was similar to the response to tail flicks (Figure 14).

However, also similarly to our study in Woo et al. (2009b), we understood the effect of varying wind speeds on vegetation might impact signal detection. In Woo & Rieucau (2013), we employed the same methodological design as in Woo & Rieucau (2012), but focused on two things: 1) the conspicuousness of aggressive (e.g., tails-flick, push-up body rock) versus submissive (e.g., slow arm wave) displays, and 2) the relative conspicuousness of social displays as wind speeds increase. When we compared the responses to aggressive and submissive displays, we found that aggressive displays were more conspicuous. This result supported our previous results that suggested that faster signals were more conspicuous. However, the additional dimension that we incorporated was the variation in wind speed. After acquiring the range of normal (Woo et al., 2009a) and uncommon wind speeds (Peters et al., 2008) on vegetation, we included the additional factor of three different wind speeds: calm, typical, and windy. Instead of superimposing generic windblown vegetation, we simulated wind speeds and windblown vegetation that matched the independent average of each treatment, and measured its consistency with AIM. Consistent with all the previous studies, we found that the signals produced by our animations were still conspicuous as wind speed increased; however, the signal did degrade slightly as simulated conditions became windier (Figure 15).

Windblown vegetation generates environmental noise with the potential to mask important visual motion. To a modest degree, faster wind conditions that increase the level of complexity in the visual field disrupt the ability for Jacky dragons to detect signals. However, all social signals are proven to be robust, retaining efficacy even under poor signaling conditions.

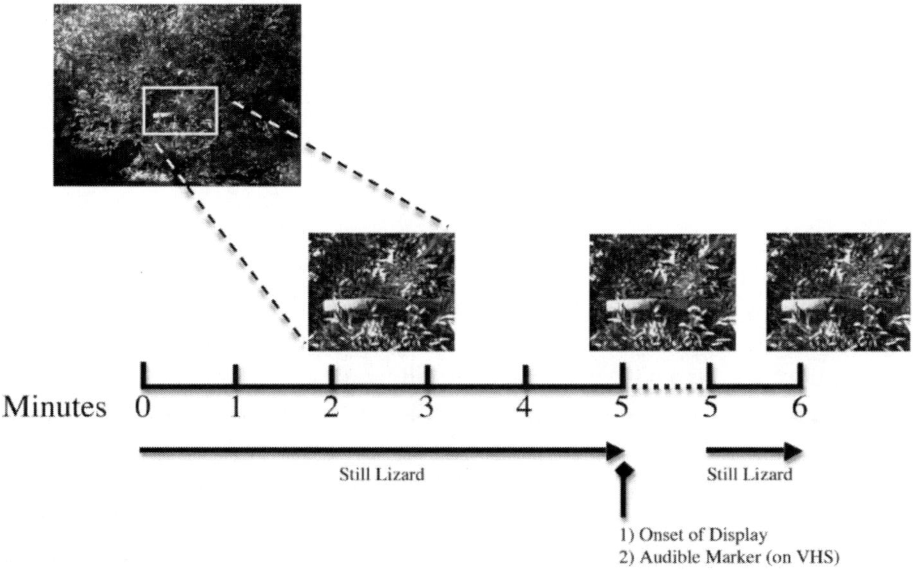

Figure 13. The entire sequence was six minutes, and during the initial five minutes, the animated lizard was perched motionless among windblown vegetation. At the five-minutes mark, the animation produced one of our programmed exemplars of displays. (Woo & Rieucau, 2012 – *Ethology*; Woo & Rieucau, 2013 – *Ethology, Ecology & Evolution*).

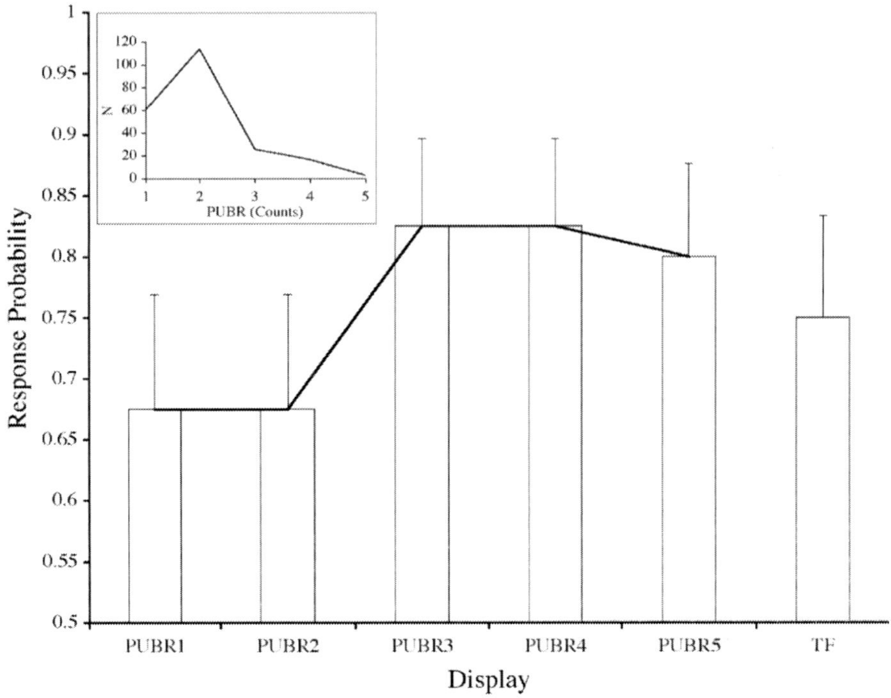

Figure 14. Optimal conspicuousness of the push-up body rocks was matched closely to the number of consecutive displays produced in a single bout (Woo & Rieucau, 2012 – *Ethology*).

Peters and Evans (2007) also employed the fully animated lizard to consider the conspicuousness of displays that incorporated the factors of distance from opponents (e.g., 1m, 3m, 9m, and 21m) and angle of presentation (e.g., frontal, diagonal, and orthogonal). Unsurprisingly, Peters & Evans (2007) found that lizards responded similarly to the angle of presentation, such as was also suggested in Woo and Rieucau (2008). However, Peters and Evans (2007) did find that signals displayed at the shortest distance were the most conspicuous, and despite a steady decease in response as length increased, displays at 21m were still conspicuous. Furthermore, as windblown vegetation is dynamic in motion, Peters (2008) suggested that Jacky lizards possibly compensate for changes to plant motion, particularly at higher speeds.

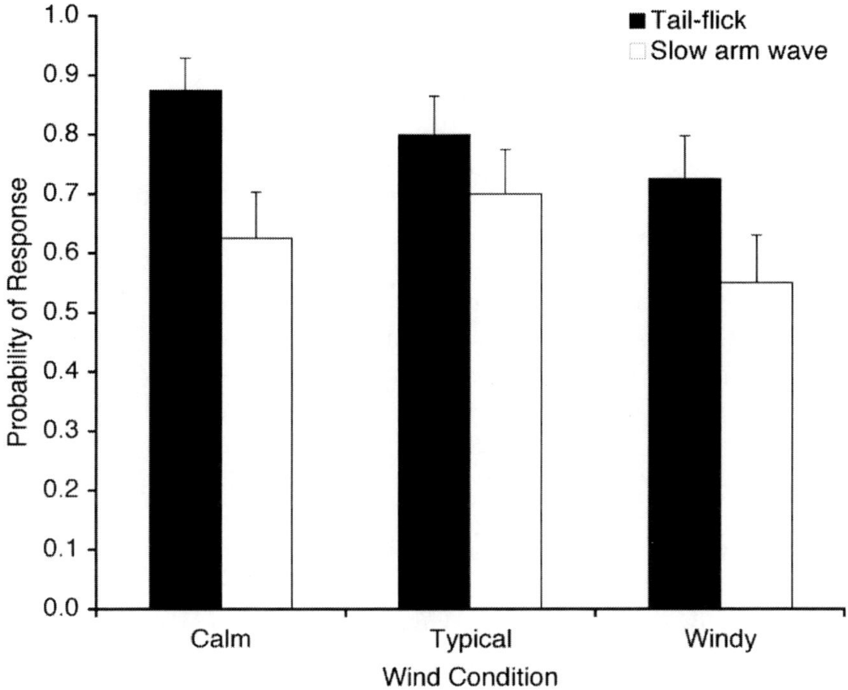

Figure 15. Both the tail-flick and slow arm wave were conspicuous across wind condition; however, signal detection became compromised as simulated wind conditions increased (Woo & Rieucau, 2013 – *Ethology, Ecology & Evolution*).

The Importance of Syntax and Morphology

If a signal is comprised of multiple components, then the sequential organization of these elements is often critical for successful communication. The particular temporal ordering of the signal units is defined as 'segmental syntax' (Marler & Peters, 1988). Syntax in human language is a function of grammatical structure with the primary aim being to build the understanding of verbal communication (Wierzbicka, 1996), and despite slight modification in the words order, the general context of ideas and overall meaning can be transmitted (Wierzbicka, 1996). In non-human communication, however, selective features on design are thought to temporally constrain the syntactical order of signals (Hailman et al., 1987; Holland et al., 2000).

It is becoming increasingly clear from the extensive study of bird song reaching various ecological contexts that recognition of complex acoustic signals is dependent upon the temporal ordering of signal units (Balaban, 1988; Brindley, 1991; Päckert et al., 1997; Pytte, 1997; Okanoya & Yamaguchi, 1997; Podos et al., 1999, Soha & Marler, 2001; Nowicki et al., 2001; Ballentine et al., 2003; Sung et al., 2005; Briefer et al., 2013). However, despite knowledge on the critical role of syntax in acoustic communication, much less is known about the analogous phenomenon of temporal arrangement of signal units in dynamic visual signals.

Tests of variation in signal syntax show that all displays in the Jacky dragon repertoire are conspicuous, regardless of order; however, social responses reveal sensitivity to temporal order. A playback experiment using high-resolution computer animations of conspecifics revealed the importance of temporal order for signal efficacy (Woo & Rieucau, 2014). Lizards were shown different life-sized simulated animations of a lizard model differing in skin texture and morphology ranging from highly natural to abnormal (Figure 16). Focal lizards perceived morphological differences of the presented models, but without influencing their subsequent behavioral responses. It has been previously shown that Jacky dragons rely on morphological cues to discriminate between familiar and unfamiliar conspecifics (Van Dyk & Evans, 2007). However, we observed that lizards responded to normal syntax with social displays, despite experimental variation in the realism of the stimulus lizard (Woo & Rieucau, 2014). Interestingly, an artificially articulated object was an effective social stimulus, so long as natural syntax (i.e., order of the motor pattern) was preserved (Figure 17). All combined these experimental results suggest that recognition of the visual display in the Jacky dragon is more dependent on sequence structure than on morphology (Van Dyk & Evans, 2007).

However, to date, some important questions remain unsolved. In particular, we noted that alterations in syntax induced different types of social responses (e.g., submissive, attack or aggressive) (Woo & Rieucau, 2014). Yet, it is unclear how Jacky dragons perceived subtle changes in information content when visual signals' syntax was violated, limiting, thus, the understanding of the functional basis of the different types of behavioral responses observed. Another important current challenge remains to ascertain the selective pressures that have shaped the particular temporal arrangement of stereotyped visual displays in this species.

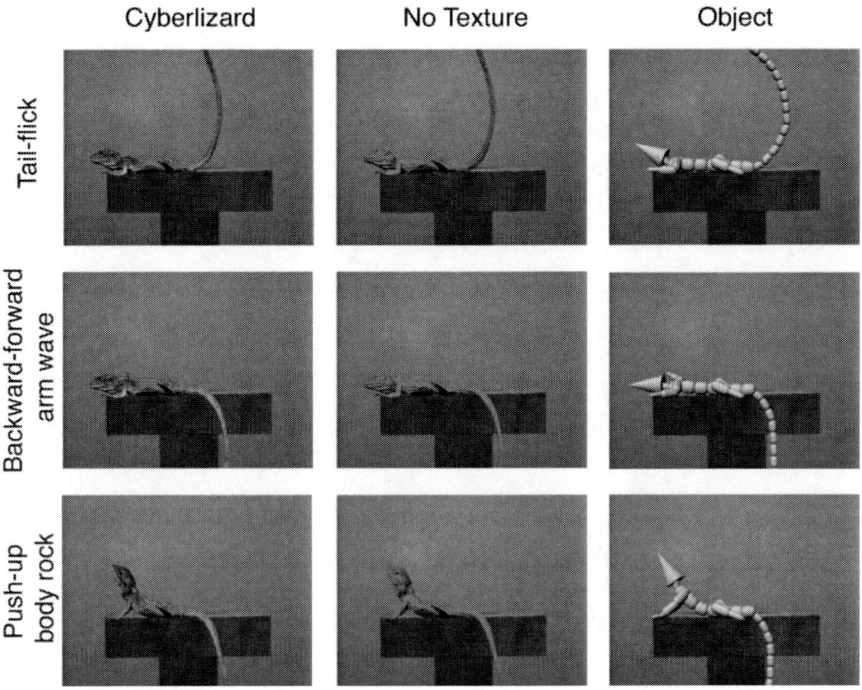

Figure 16. Photographic comparison of the three animated models (e.g., realistic, no texture, and object) engaged in normal syntax (Woo & Rieucau, 2014 – *Acta Ethologica*).

Interactive Video Playback Experiments Using Computer-Generated Animations

Interactivity has been an important critical challenge to overcome towards the development of highly realistic video playback experiments in the study of animal visual communication. As communication relies upon dynamic and somehow complex interactions between individuals, a major deficit of the conventional video playback technique is its incapacity to accurately reproduce the temporal continuity of social scenes. Interactive video playbacks provide researchers with an efficient means to enhance the realistic level of staged social interactions. In the Jacky dragon, Ord & Evans (2002) carried out one of the first experiments using interactive video sequences of visual displays, showing that the addition of natural reciprocal responses exhibited by a simulated animated conspecifc affect the responses of live observers, whereas non-interactive video sequences reduce or even suppress these responses. Hence, their results provided supporting evidence of the utility of the interactive video playback approach in the exploration of the functional basis of Jacky dragon's complex visual communication.

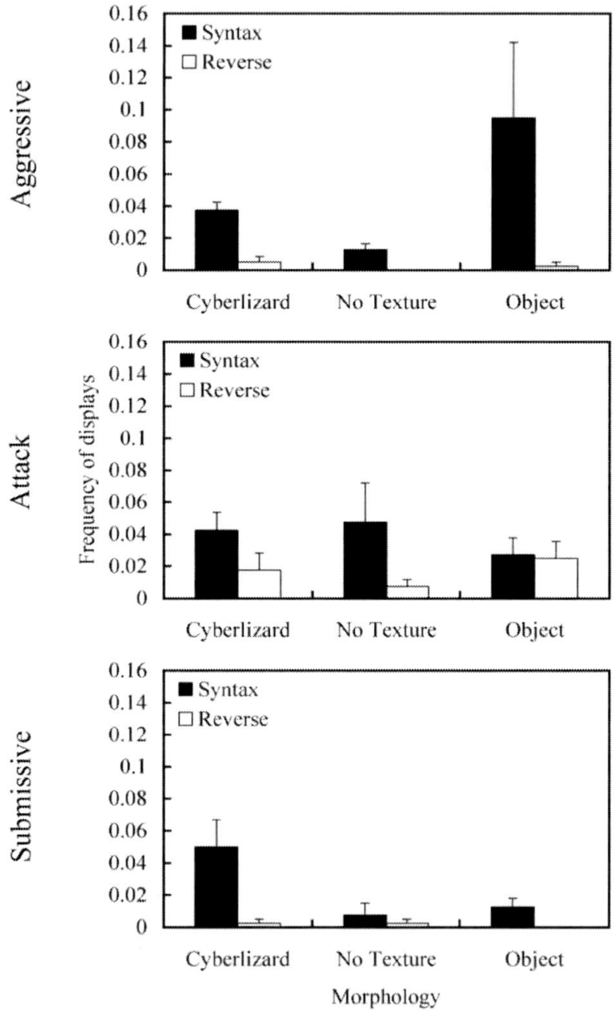

Figure 17. Across the three clustered response measures (e.g., aggressive, attack, and submissive), lizards responded more significantly to normal syntax for all three models, as opposed to the reversed order (Woo & Rieucau, 2014 - *Acta Ethologica*).

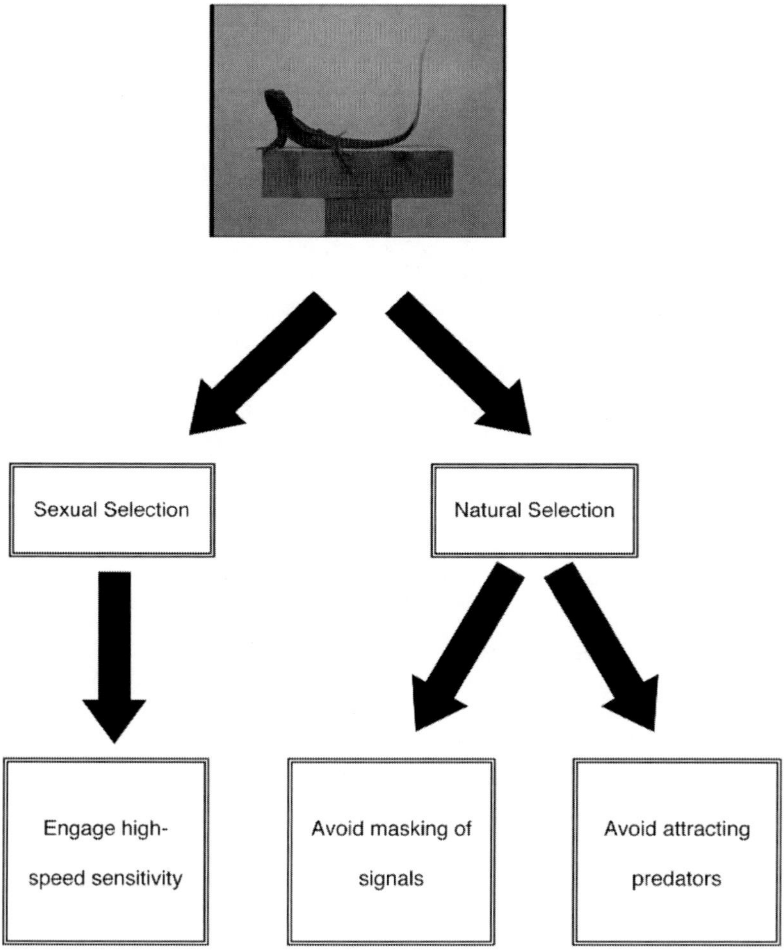

Figure 18. Sexual and natural selection pressures have shaped the design of the high-speed tail-flick. High-speed motion is necessary to stimulate the receiver sensory system, and draw attention to the signaler. The signal must also overcome the masking of windblown vegetation; however, high-speed motion also has the potential to attract aerial and ground predators.

The potential of this approach shows in the context of opponent assessment of fighting ability in the Jacky dragon where interactive video playback of computer-generated animations have been successfully used to investigate the inherent functions of the various signals employed by lizards during male-male encounters (Ord & Evans, 2002; Van Dyk & Evans, 2008). By controlling the behavioral signals produced by a computer-generated animated opponent in a way that matched or deviated from the strategic responses of live lizards, Van Dyk & Evans (2008) found that males' decisions to initiate or not an attack were based on the characteristics of the sequence of visual signals displayed by the simulated opponent (submissive or aggressive displays). This technique also can be used to further investigate the role of visual signals syntax on Jacky dragons' strategic decisions during interactions with social partners.

Since playback permits to precise control over the sequential displayed responses of a simulated companions relative to the behaviors of an observer in a 'moment-to-moment' fashion (Ord & Evans, 2002), the integrative video playback technique that uses computer-generated animations is particularly well-suited for the experimental study of the Jacky dragon's visual communication and its dynamic stereotyped repertoire in a broad ranges of ecological contexts such as mate selection or territoriality.

CONCLUSION

The application of computer-generated animations has provided another avenue of investigation for examining questions not only in human biological motion, but also in animal communication and signal design. This method has been successfully tested (i.e., signal conspicuousness: Peters & Evans, 2003a,b; active space: Peters & Evans, 2007; interactive playback and opponent assessment: Van Dyk & Evans, 2008; visual syntax: Woo & Rieucau, 2014), and the results that have provided support that live animals clearly interact with animations.

The use computer-generated animations allow researchers to explore questions in signal design beyond examining staged interactions. Previous studies have made minor manipulations to morphology, such as changing the length of swordtails (Morris et al., 2005; Wong & Rosenthal, 2006), manipulating the number of stripes in zebrafish (Rosenthal & Ryan, 2005), varying the symmetry of pelvic spine in sticklebacks (Mazzi et al., 2003), and altering the placement of the large anterior median eyes in jumping spiders (Harland & Jackson, 2002). However, we can further manipulate these morphological attributes to address theoretical principles in signal design and function.

The use of video playback and computer animations reduces the use of live animals while controlling for behavioral variables, such as morphology and signal structure. In the application of this technique, it is critical in understanding the nature of specific visual signals in terms of their mechanistic function, perception, and evolution.

Moreover, the studies on Jacky dragons that have employed computer-generated animations to stage interactions also highlight the constraints that limit the selective forces on signal design. Here, in our conclusion, we propose how two selective pressures (e.g., sensory and environmental) have shaped the evolution of signal design in the Jacky dragon.

Sensory: Sexual Selection versus Natural Selection Pressures

Selection for signal conspicuousness is likely constrained by an interaction between sexual and natural selection (Figure 18). The tail-flick is a high-speed display used to attract the attention of neighboring males or potential mates (Peters & Evans, 2003a). It is also critical that signals successfully attract the attention of conspecifics when they are displayed against the visual complexity of windblown movement in different habitat structures (Ord et al., 2002; Peters et al., 2008) and across varying distances (Peters & Evans, 2007). However, it is also likely that high-speed motion causes the signaler more conspicuous to aerial predators, such as raptors. To avoid attracting the attention of predators, Jacky dragons must produce fast signals to increase efficacy, but also limit the overall length of the display.

In a comparative study across all other agamid lizards, Stuart-Fox and Ord (2004) suggested that natural selection is constrained by the evolution of elaborate ornaments and driving them to be inconspicuous to predators. Consequently, dichromatism for inconspicuous body color to avoid exposure to visual predators is driven by sexual selection (Stuart-Fox & Ord, 2004). We suggest that in the Jacky dragon, sexual selection and natural selection have influenced the evolution of a movement-based display in an analogous fashion.

Environmental: Improve Efficacy to Overcome Signal Masking

The efficacy of visual displays is constrained by the movement of windblown vegetation. To overcome the masking effect of this type of environmental noise, signals must evolve a design component to compensate for complex visual noise; moreover, the receiver visual system must then develop a mechanism to enhance the detection of these signals (Peters &

Evans, 2003b). The effect of ambient noise on signal detection has recently been given much attention in the literature, especially in acoustic signals affected by urban noise (Patricelli & Blickley, 2006; Slabbekoorn & den Boer-Visser, 2006; Wood & Yezerinac, 2006). For example, great tits (*Parus major*) increase the frequency pitch of their songs to prevent masking from human-induced noise (Slabbekoorn & Peet, 2003). Similarly, house finches (*Carpodacus mexicanus*) increase singing rates to overcome ambient noise in urbanised areas (Fernández-Jurcic et al., 2005). In the visual system, the dewlap display of *Anolis* lizards is dependent on rapid acceleration to enhance signal detection and efficacy (Fleishman, 1992). *Anolis* lizards speed up the rate of the dewlap display when wind conditions increase (Ord et al., 2007). However, in Jacky dragons, signal efficacy is principally dependent on duration, as seen in both the tail-flick (Peters & Evans, 2003b; Peters et al., 2007) and the push-up body rock (Woo & Rieucau, 2012). For the *Anolis* lizard and Jacky dragon visual systems, two different designs have evolved to solve the same problem of signal detection when displays are produced in complex noise.

To increase signal efficacy, Jacky dragons can increase the number of push-up body rock displays (Figure 19); however, both energetic and predation costs likely limit the maximum number of push-up body rocks in a bout. For example, side-blotched lizards (*Uta stansburiana*) that have been experimentally fatigued are shown to display less, thus reducing the intensity of displays and success in staged contests (Brandt, 2003; Brandt & Allen, 2004). Furthermore, lengthy signal production also increases the conspicuousness of the display to potential predators (Yasukawa, 1989). Further research to explore these issues may provide a better understanding of the trade-off between signal efficacy and costs in the Jacky dragon.

The visual signals of Jacky dragons are clearly designed to be robust against a range of natural wind conditions. Jacky lizards are also commonly found across a variable range of habitat (Harlow & Taylor, 2000). In sympatric bird species, it is suggested that neighbors have distinctive acoustic spaces (Nelson & Marler, 1990). A comparison between regional populations, such as in Asian nations where sympatric agamid lizards occur, may identify whether displays have evolved in local populations as a constraint defined by visual signal spaces. As observed in the dewlap display of *Anolis* lizards (Blotch & Irschick, 2006), it would also be of interest to compare display characteristics to those of other local populations, and whether similarities or differences may be attributed to local vegetation structure or geography.

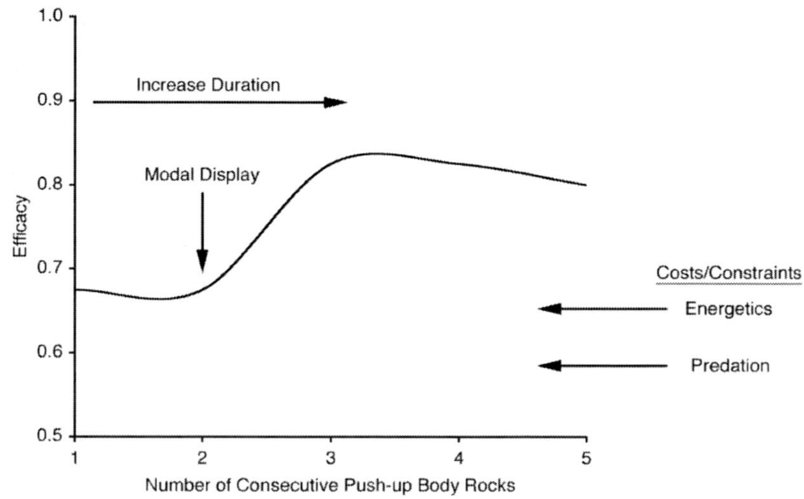

Figure 19. Display duration increases signal efficacy. However, the push-up body rock is not open-ended, and is limited to five consecutive displays. Signal duration is likely constrained by two factors: energetic constraints and predation. The production of lengthy displays may expend energetic resources necessary for agonistic or mating interactions. In addition, lengthy displays also increase the signaler's conspicuousness.

Conspicuousness of a Broadcast Signal Is a Cue for Recognition

Free from the competitive pressures of sympatric agamid lizards, Jacky dragon signals maintain a strict salience hierarchy. The tail-flick is the most conspicuous display. As an archetypal broadcast signal, the tail-flick comprises of all the signal design components necessary for it to be highly conspicuous, such as high-speed motion, large sweep area (Peters & Evans, 2003a), and greatest amplitude structure (Peters & Evans, 2003a). Likely constrained by energetic costs and the potential to attract predators, the tail-flick evolved a signal design component, such that the duration of the signal increases efficacy (Peters & Evans, 2003b). Thus, the tail-flick is exceptionally designed to attract the attention of potential receivers; however, this display does not contain any social information. Once the tail-flick is conspicuous to receivers, another motor pattern that contains information about the signaler is required. As a result, the evolution of additional signals, such as the push-up body rock, follows the broadcast display. The recognition of signals that contain social information is hence a secondary consequence of conspicuousness efficacy (Table 4). Once invariant, the tail-flick display is a potential cue for recognition of signal salience order.

After receiving information about an opponent (Table 4), the receiver must then assess status, and select a tactic in which to respond. The receiver may choose to reciprocate with an aggressive signal and escalate the conflict, or respond with an appeasement display to reduce aggression and prevent injury (Neat et al., 1998). Ord et al. (2001) suggested that signal complexity evolved to improve opponent assessment that was essentially a consequence of male-male competition. As a result, opponent assessment likely requires a degree of higher order cognitive decision-making to assess whether the benefits of interacting and potentially winning outweigh the costs of defeat (Enquist & Leimar, 1983; Enquist & Jakobson, 1986). Previous studies show that Jacky dragons can assess opponent status relatively quickly using morphological cues and display rate (Van Dyk et al., 2007). However, it is not known how opponent assessment is influenced by changes to the fine structure of displays, such as manipulating the speed, amplitude, and duration of the signal components. Thus, future work may also identify other necessary features of the Jacky dragon signal structure that are critical for opponent assessment.

Table 4. A salience hierarchy exists where the tail-flick is the most conspicuous, and this is followed by the push-up body rock. The tail-flick must have evolved first as an assertive signal and the push-up body rock likely evolved as a secondary consequence of the tail-flick. However, the push-up display contains more information than a broadcast signal and requires a mechanism for recognition. Consequently, a level of cognitive assessment is required before the receiver elects to respond to the signaler

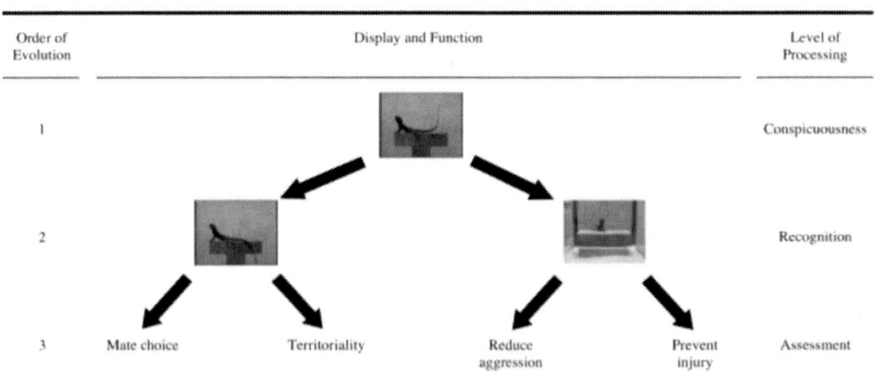

Further comparative studies may elucidate the link between the evolution of conspicuously designed signals and the recognition mechanisms required for cognitive assessment of conspecifics. To date, studies only adopt a comparative approach to examine the link between the evolution and signal characteristics, and typically measure the trait as simply present or absent across species (Ord et al., 2001; Ord et al., 2002; Stuart-Fox & Ord, 2004). There is, hence, a need for analogous studies to devise a graded scale to measure signal traits, such that researchers are able to reconstruct evolutionary events more systematically. Thus, developing a graded scale of signal traits may identify whether the evolution of sensory processes indeed precedes the ability to recognize patterns, and at what evolutionary juncture that this link may have possibly been established.

ACKNOWLEDGMENTS

KLW was funded by an Australian Research Council Discovery Grant and a Macquarie University CISAB Postgraduate Award. GR was supported by a postdoctoral fellowship from the Institute of Marine Research and by the Norwegian Research Council (grant 204229/F20). Permission to collect lizards was granted by the New South Wales National Parks and Wildlife Service (License No. S11024). Approval for ethical research and husbandry guidelines were granted by the Macquarie University Animal Ethics Committee (Animal Research Protocol No. 2003/014). KLW would like to particularly thank C. S. Evans, D. Burke, R. A. Peters, P. Radhakrishnan, R. R. Woo, and K. N. Woo. KLW and GR would also like to thank the Junior Team C.T.F.D.

REFERENCES

Basolo, A. L. (1990). Female preference predates the evolution of the sword in swordtail fish. *Science*, *250*, 808-810.

Basolo, A. L. (1995a). A further examination of pre-existing bias favouring a sword in the genus *Xiphophorus*. *Animal Behavaiour*, *50*, 365-375.

Basolo, A. L. (1995b). Phylogenetic evidence for the role of a pre-existing bias in sexual selection. *Proceedings of the Royal Academy of London Series B*, *259*, 307-311.

Bels, V. (1986). Analysis of the display-action-pattern of *Anolis chlorocyanus* (Sauria: Iguanidae). *Copeia, 1986*, 963-970.

Bennett, A. T. D. & Cuthill, I. C. (1994). Ultraviolet vision in birds: What is its function? *Vision Research, 34*, 1471-1478.

Balaban, E. (1988). Bird song syntax: learned intraspecific variation is meaningful. *Proceedings of National Academy of Sciences USA, 85*, 3657-3660.

Ballentine, B., Badyaev, A. & Hill, G. E. (2003). Changes in song complexity correspond to periods of female fertility in blue grosbeaks (*Guiraca caerulea*). *Ethology, 109*, 55-66.

Bischof, W. F., Reid, S. L., Wylie, D. W. & Spetch, M. L. (1999). Perception of coherent motion in random dot displays by pigeons and humans. *Perception & Psychophysics, 61*, 1089-1101.

Blotch, N. & Irschick, D. J. (2006). An analysis of inter-population divergence in visual display behavior of the green anole lizard (*Anolis carolinensis*). *Ethology, 112*, 370-378.

Blumstein, D. T. & Armitage, K. B. (1997). Does sociality drive the evolution of communicative complexity? A comparative test with ground-dwelling sciurid alarm calls. *American Naturalist, 150*, 179-200.

Bradbury, J. W. & Vehrencamp, S. L. (1998). *Principles of animal communication*. Sunderland: Sinauer Associates, Inc.

Brandt, Y. (2003). Lizard threat display handicaps endurance. *Proceedings of the Royal Society of London Series B, 270*, 1061-1068.

Brandt, Y. & Allen, J. R. (2004). Persistence of individually distinctive display patterns in fatigued side-blotched lizards (*Uta stansburiana*). *Behavioral Ecology and Sociobiology, 55*, 257-265.

Briefer, E. F., Rybak, F. & Aubin, T. (2013). Does true syntax or simple auditory object support the role of skylark song dialect? *Animal Behaviour, 86*, 1131-1137.

Brindley, E. L. (1991). Response of European robins to playback of song: neighbour recognition and overlapping. *Animal Behaviour, 41*, 503-512.

Burghardt G. M. (1977). Learning processes in reptiles. In: Gans C, Tinkle DW (Eds.) In C. Gans, & D. W. Tinkle (Eds.), *Biology of the reptilia* (pp. 555–681). New York: Academic Press.

Burnat, K., Vandenbussche, E. & Zernicki, B. (2002). Global motion detection is impaired in cats deprived of pattern vision. *Behavioural Brain Research, 134*, 59–65.

Carpenter, C. C. Badham, J. A. & Kimble, B. (1970). Behavior patterns of three species of *Amphibolurus* (Agamidae). *Copeia, 1970*, 497-505.

Carpenter, C. C. & Ferguson, G. W. (1977). Variation and evolution of stereotyped behavior in reptiles. In C. Gans, & D. W. Tinkle (Eds.), *Biology of the reptilia* (pp. 335-554). New York: Academic Press.

Christy, J. H. (1988). Pillar function in the fiddler crab, *Uca beebei.* (II) Competitive courtship signalling. *Ethology*, *78*, 113-128.

Christy, J. H. (1995). Mimicry, mate choice, and the sensory bias trap. *American Naturalist*, *146*, 171-181.

Clark, D. & Uetz, G. W. (1990). Video image recognition by the jumping spider, *Maevia inclemens* (Araneae: Salticidae). *Animal Behaviour*, *40*, 884-890.

Clifford, C. W. G. & Ibbotson, M. R. (2003). Fundamental mechanisms of visual motion detection: Models, cells, and functions. *Progress in Neurobiology*, *68*, 409-437.

Cook, R. G. & Katz, J. S. (1999). Dynamic object perception by pigeons. *Journal of Experimental Psychology: Animal Behavior Processes*, *l 25*, 194-210.

Cook, R. G., Shaw, R. & Blaisdell, A. P. (2001). Dynamic object perception by pigeons: Discrimination of action in video presentations. *Animal Cognition*, *4*, 137-146.

Curran, W. & Braddick, O. J. (2000). Speed and direction of locally-paired dot patterns. *Vision research*, *40*, 2115-2124.

Davson, H. (1990). *Physiology of the eye.* New York: Pergamon.

D'Eath, R. B. (1998). Can video images imitate real stimuli in animal behaviour experiments? *Biological Reviews*, *73*, 267-292.

Decourcy, K. R. & Jenssen, T. A. (1994). Structure and use of male territorial headbob signals by the lizard *Anolis carolinensis*. *Animal Behaviour*, *47*, 251-262.

Endler, J. A. (1987). Predation, light intensity and courtship behaviour in *Poecilia reticulata* (Pisces: Poeciliidae). *Animal Behaviour*, *30*, 1376-1382.

Endler, J. A. (1992). Signals, signal conditions, and the direction of evolution. *American Naturalist*, *139*, S125-153.

Endler, J. A. (1993). Some general comments on the evolution and design of animal communication systems. *Philosophical Transactions of the Royal Society of London B*, *340*, 215-225.

Endler, J. A. & Basolo, A. L. (1998). Sensory ecology, receiver biases and sexual selection. *Trends in Ecology & Evolution*, *13*, 415-420.

Enquist, M. & Jakobson, S. (1986). Decision making and assessment in the fighting behaviour of *Nannacara anomala* (Cichlidae, Pisces). *Ethology, 72*, 143-153.

Enquist, M. & Leimar, O. (1983). Evolution of fighting behaviour: Decision rules and assessment of relative strength. *Journal of Theoretical Biology, 102*, 387-410.

Evans, C. S. & Marler, P. (1991). On the use of video images as social stimuli in birds: Audience effects on alarm calling. *Animal Behaviour, 41*, 17–26.

Fernández-Juricic, E., Poston, R., De Collibus, K., Morgan, T., Bastain, B., Martin, C., Jones, K. & Treminio, R. (2005). Microhabitat selection and singing behavior patterns of male house finches (*Carpodacus mexicanus*) in urban parks in a heavily urbanized landscape in the Western U.S. *Urban Habitats, 3*, 49-69.

Fleishman, L. J. (1988a). Sensory and environmental influences on display form in *Anolis auratus*, a grass anole from Panama. *Behavioral Ecology and Sociobiology, 22*, 309-316.

Fleishman, L. J. (1988b). Sensory influences on physical design of a visual display. *Animal Behaviour, 36*, 1420-1424.

Fleishman, L. J. (1992). The influence of the sensory system and the environment on motion patterns in the visual displays of anoline lizards and other vertebrates. *American Naturalist, 139*, S36-S61.

Fleishman, L. J. & Endler, J. A. (2000). Some comments on visual perception and the use of video playback in animal behavior studies. *Acta Ethologica, 3*, 15-27.

Fleishman, L. J., McClintock, W. J., D'Eath, R. B., Brainards, D. H. & Endler, J. A. (1998). Colour perception and the use of video playback experiments in animal behaviour. *Animal Behaviour, 56*, 1035-1040.

Fleishman, L. J. & Persons, M. (2001). The influence of stimulus and background colour on signal visibility in the lizard *Anolis cristatellus. Journal of Experimental Biology, 204*, 1559-1575.

Gatesy, S. M., Middleton, K. M., Jenkins F. A., Jr. & Shubin, N. H. (1999). Three-dimensional preservation of foot movements in Triassic theropod dinosaurs. *Nature, 399*, 141-144.

Gibbons, J. R. H. (1979). The hind leg pushup display of the *Amphibolurus decresii* species complex (Lacertilia: Amagidae). *Copeia, 1979*, 29-40.

Goto, K., Lea, S. E. G. & Dittrich, W. H. (2002). Discrimination of intentional and random motion paths by pigeons. *Animal Cognition, 5*, 119–127.

Guilford, T. & Dawkins, M. S. (1991). Receiver psychology and the evolution of animal signals. *Animal Behaviour, 42*, 1-14.

Hailman, J. P., Ficken, M. S. & Ficken, R. W. (1987). Constraints on the structure of combinatorial "chick-a-dee" calls. *Ethology*, *75*, 62-80.

Harland, D. P. & Jackson, R. R. (2002). Influence of cues from the anterior medial eyes of virtual prey on *Portia fimbriata*, an araneophagic jumping spider. *Journal of Experimental Biology*, *205*, 1861-1868.

Harlow, P. S. & Taylor, J. E. (2000). Reproductive ecology of the Jacky dragon (*Amphibolurus muricatus*): An agamid lizard with temperature-dependent sex determination. *Austral Ecology*, *25*, 640-652.

Holland, J., Dabelsteen, T. & Paris, A. L. (2000). Coding in the song of the wren: importance of rhythmicity, syntax and element structure. *Animal Behaviour*, *60*, 463-470.

Jenssen, T. A. (1977). Evolution of anoline lizard display behaviour. *American Zoologist*, *17*, 203-215.

Jenssen, T. A. & Swenson, B. (1974). An ecological correlate of critical flicker-fusion frequencies for some *Anolis* lizards. *Vision Research*, *14*, 965-970.

Johnstone, R. A. (1994). Female preference for symmetrical males as a by-product of sexual selection for mate recognition. *Nature*, *372*, 172-175.

Johnstone, R. A. (1995). Sexual selection, honest advertisement and the handicap principle: Reviewing the evidence. *Biological Reviews*, *70*, 1-65.

Kingstone, A. & Bischof, W. F. (1999). Perceptual grouping and motion coherence in visual search. *Psychological Science*, *10*, 151-156.

Kiorpes, L. & Movshon, J. A. (2004). Development of sensitivity to visual motion in macaque monkeys. *Visual Neuroscience*, *21*, 851-859.

Kordic-Brown, A. & Nicoletto, P. F. (1997). Repeatability of female choice in the guppy: Response to live and videotaped males. *Animal Behaviour*, *54*, 369-376.

Künzler, R. & Bakker, T. C. M. (1998). Computer animations as a tool in the study of mating preferences. *Behaviour*, *135*, 1137-1159.

Leal, M. & Rodríguez-Robles, J. A. (1997). Signalling displays during predator-prey interactions in a Puerto Rican anole, *Anolis cristatellus*. *Animal Behaviour*, *54*, 1147-1154.

Liu, Z. & Weinshall, D. (2000). Mechanisms of generalization in perceptual learning. *Vision Research*, *40*, 97-109.

Lythgoe, J. N. (1979). *The ecology of vision.* Oxford: Claredon Press.

Maddocks, S. A., Goldsmith, A. R. & Cuthill, I. C. (2001). The influence of flicker rate on plasma corticosterone levels of European starlings, *Sturnus vulgaris*. *General and Comparative Endocrinology*, *124*, 315-320.

Marchetti, K. (1993). Dark habitats and bright birds illustrate the role of environment in species divergence. *Nature*, 362, 149-152.

Marler, P. & Peters, S. (1988). The role of song phonology and syntax in vocal learning preferences in the song sparrow, *Melospiza melodia. Ethology*, 77, 125-149.

Matthews, N., Liu, Z., Geesaman, B. J. & Qian, N. (1999). Perceptual learning on orientation and direction discrimination. *Vision Research*, *39*, 3692-3701.

Mazzi, D., Künzler, R. & Bakker, T. C. (2003). Female preference for symmetry in computer-animated three-spined sticklebacks, *Gasterosteus aculeatus*. *Behavioral Ecology and Sociobiology*, *54*, 156-161.

McKinnon, J. S. & McPhail, J. D. (1996). Male aggression and colour in divergent populations of the threespine stickleback Experiment with animations. *Canadian Journal of Zoology*, *74*, 1727-1733.

Morris, M. R., Moretz, J. A., Farley, K. & Nicoletto, P. (2005). The role of sexual selection in the loss of sexually selected traits in the swordtail fish *Xiphophorus continens*. *Animal Behaviour*, *69*, 1415-1424.

Morton, E. S. (1975). Ecological sources of selection on avian sounds. *American Naturalist*, *109*, 17-34.

Neat, F. C., Taylor, A. C. & Huntingford, F. A. (1998). Proximate costs of fighting in male cichlid fish: The role of injuries and energy metabolism. *Animal Behaviour*, *55*, 875-882.

Nelson, D. A. & Marler, P. (1990). The perception of birdsong and an ecological concept of signal space. In W. C. Stebbins & M. A. Berkley (Eds.), *Comparative Perception – Volume II: Complex Signals* (pp. 443-478). New York: John Wiley & Sons, Inc.

Newsome, W. T. & Paré, E. B. (1988). A selective impairment of motion perception following lesions of the middle temporal visual area (MT). *Journal of Neuroscience*, *8*, 2201-2211.

Nowak, L. M. & Green, D. G. (1983). Flicker fusion characteristics of rod receptors in the toad. *Vision Research*, *23*, 845-849.

Nowicki, S., Searcy, W. A., Hughes, M. & Podos, J. (2001). The evolution of bird song: male and female response to song innovation in swamp sparrows. *Animal Behaviour*, *62*, 1189-1195.

Okanoya, K. & Yamaguchi, A. (1997). Adult bengalese finches (*Lonchura striata* var. *domestica*) require real-time auditory feedback to produce normal song syntax. *Journal of Neurobiology*, *33*, 343-356.

Oliveira, R. F., Rosenthal, G. G., Schlupp, I., McGregor, P. K., Cuthill, I. C., Endler, J. A., Fleishman, L. J., Zeil, J., Barata, E., Burford, F., Gonçalves,

D., Haley, M., Jakobsson, S., Jennions, M. D., Körner, K. E., Lindström, L., Peake, T., Pilastro, A., Pope, D. S., Roberts, S. G. B., Rowe, C., Smith, J. & Waas, J. R. (2000). Considerations on the use of video playbacks as visual stimuli: The Lisbon workshop consensus. *Acta Ethologica*, *3*, 61-65.

Ord, T. J., Blumstein, D. T. & Evans, C. S. (2001). Intra-sexual selection predicts the evolution of signal complexity in lizards. *Proceedings of the Royal Society of London Series B*, *268*, 737-744.

Ord, T. J., Blumstein, D. T. & Evans, C. S. (2002). Ecology and signal evolution in lizards. *Biological Journal of the Linnean Society*, *77*, 127-148.

Ord, T. J., Peters, R. A., Clucas, B. & Stamps, J. A. (2007). Lizards speed up visual displays in noisy motion habitats. *Proceedings of the Royal Society of London Series B*, *274*, *1075-1062.*

Ord, T. J. & Evans, C. S. (2002). Interactive video playback and opponent assessment in lizards. *Behavioural Processes*, *59*, 55-65.

Ord, T. J., Peters, R. A., Evans, C. S. & Taylor, A. J. (2002). Digital video playback and visual communication in lizards. *Animal Behaviour*, *63*, 879-890.

Päckert, M., Martens, J., Kosuch, J., Nazarenko, A. A. & Veith, M. (2003). Phylogenetic signal in the song of crests and kinglets (Aves: *Regulus*). *Evolution*, *57*, 616-629.

Patricelli, G. L. & Blickley, J. L. (2006). Avian communication in urban noise: Causes and consequences of vocal adjustment. *Auk*, *123*, 639-649.

Peters, R. A. (2008). Environmental motion delays the detection of movement-based signals. *Biology letters*, *4*, 2-5.

Peters, R. A., Clifford, C. W. G. & Evans, C. S. (2002). Measuring the structure of dynamic visual signals. *Animal Behaviour*, *64*, 131-146.

Peters R. A. & Evans, C. S. (2003a). Introductory tail-flick of the Jacky dragon visual display: Signal efficacy depends upon duration. *Journal of Experimental Biology*, *206*, 4293-4307.

Peters, R. A. & Evans, C. S. (2003b). Design of the Jacky dragon visual display: Signal and noise characteristics in a complex moving environment. *Journal of Comparative Physiology A*, *189*, 447-459.

Peters, R. A. & Evans, C. S. (2007). Active space of a movement-based signal: response to the Jacky dragon (*Amphibolurus muricatus*) display is sensitive to distance, but independent of orientation. *Journal of Experimental Biology*, *210*, 395-402.

Peters, R., Hemmi, J. & Zeil, J. (2008). Image motion environments: background noise for movement-based animal signals. *Journal of Comparative Physiology A*, *194*, 441-456.

Peters, R. A., Hemmi, J. M. & Zeil, J. (2007). Signaling against the wind: Modifying motion-signal structure in response to increased noise. *Current Biology*, *17*, 630-636.

Peters, R. A. & Ord, T. J. (2003). Display of the Jacky dragon, *Amphibolurus muricatus*, (Lacertilia: Agamidae), to intruders: A semi-Markovian process. *Austral Ecology*, *28*, 499-506.

Podos, J., Nowicki, S. & Peters, S. (1999). Permissiveness in the learning and development of song syntax in swamp sparrows. *Animal Behaviour*, *58*, 93-103.

Pytte, C. (1997). Song organization of house finches at the edge of an expanding range. *Condor*, *58*, 93-103

Rosenthal, G. G. (2000). Design considerations and techniques for constructing video stimuli. *Acta Ethologica*, *3*, 49-54.

Rosenthal, G. G. & Evans, C. S. (1998). Female preference for swords in Xiphophorus helleri reflects a bias for large apparent size. *Proceedings of the National Academy of Sciences*, *95*, 4431-4436.

Rosenthal, G. G. & Ryan, M. J. (2005). Assortative preferences for stripes in danios. *Animal Behaviour*, *70*, 1063-1066.

Rowe, C. (1999). Receiver psychology and the evolution of multicomponent signal. *Animal Behaviour*, *58*, 921-931.

Rowland, W. J. (1995). Do female stickleback care about male courtship vigour? Manipulation of display tempo using video playback. *Behaviour*, *132*, 951-961.

Rowland, W. J., Bolyard, K. J., Jenkins, J. J. & Fowler, J. (1995). Video playback experiments on stickleback mate choice: Female motivation and attentiveness to male colour cues. *Animal Behaviour*, *49*, 1559-1567.

Ryan, M. J. (1985). *The túngara frog: A study in sexual selection and communication*. Chicago: University of Chicago Press.

Ryan, M. J. & Brenowitz, E. A. (1985). The role of body size, phylogeny, and ambient noise in the evolution of bird song. *American Naturalist*, *126*, 87-100.

Ryan, M. J., Cocroft, R. B. & Wilczynski, W. (1990). The role of environmental selection in intraspecific divergence of mate recognition signals in the cricket frog, *Acris cepitans*. *Evolution*, *44*, 1869-1872.

Ryan, M. J. & Kime, N. M. (2003). Selection on long-distance acoustic signals. In A. M. Simmons, A. N. Popper, & R. R. Fay (eds.), *Acoustic communication* (pp. 225-273). New York: Springer.

Ryan, M. J. & Rand, A. S. (1993). Sexual selection and signal evolution: The ghost of biases past. *Philosophical Transactions of the Royal Society of London Series B*, *340*, 187-195.

Scase, M. O., Braddick, O. J. & Raymond, J. E. (1996). What is noise for the motion system? *Vision Research*, *36*, 2579-2586.

Sekular, R. & Blake, R. (1994). *Perception*. McGraw-Hill, New York.

Slabbekoorn, H. & den Boer-Visser, A. (2006). Cities change the songs of birds. *Current Biology*, *16*, 2326-2331.

Slabbekoorn, H. & Peet, M. (2006). Birds sing at a higher pitch in urban noise. *Nature*, *242*, 267.

Soha, J. A. & Marler, P. (2001). Vocal syntax development in the white-crowned sparrow (*Zonotrichia leucophyrus*). *Journal of Comparative Psychology*, *115*, 172-180

Srinivasan, M. V., Zhang, S., Lehrer, M. & Collett, T. S. (1996). Honeybee navigation enroute to the goal: Visual flight control and odometry. *Journal of Experimental Biology*, *199*, 237-244.

Stuart-Fox, D. M. & Ord, T. J. (2004). Sexual selection, natural selection and the evolution of dimorphic coloration and ornamentation in agamid lizards. *Proceedings of the Royal Society of London Series B*, *271*, 2249-2255.

Sung, H.-C., Miller, E. H. & Flemming, S. P. (2005). Breeding vocalizations of the piping plover (*Charadrius melodus*): structure, diversity, and repertoire organization. *Canadian Journal of Zoology*, *83*, 579-595.

Tokarz, R. R., Patterson, A. V. & McMann, S. (2003). Laboratory and field tests of the functional significance of the male's dewlap in the lizard *Anolis sagrei*. *Copeia*, *2003*, 502-511.

Van Dyk, D. A. & Evans, C. S. (2007). Familiar-unfamiliar discrimination based on visual cues in the Jacky dragon, *Amphibolurus muricatus*. *Animal Behaviour*, *74*, 33-44.

Van Dyk, D. A. & Evans, C. S. (2008). Opponent assessment in lizards: examining the effect of aggressive and submissive signals. *Behavorial Ecology*, *19*, 895-901.

Van Dyk, D. A., Taylor, A. J. & Evans, C. S. (2007). Assessment of repeated displays: a test of possible mechanisms. *Journal of Experimental Biology*, *210*, 3027-3035.

Verghese, P. & McKee, S. P. (2004). Visual search in clutter. *Vision Research*, *44*, 1217-1225.

Vreven, D. & Verghese, P. (2002). Integration of speed signals in the direction of motion. *Perception & Psychophysics*, *64*, 996-1007.

Watanabe, S. & Troje, N. F. (2006). Towards a "virtual pigeon": A new technique for investigating avian social perception. *Animal Cognition*, *9*, 271-279.

Warner, D. A. & Shine, R. (2005). The adaptive significance of temperature-dependent sex determination: Experimental tests with a short-lived lizard. *Evolution*, *59*, 2209-2221.

Warner, D. A., Woo, K. L., Van Dyk, D. A., Evans, C. S. & Shine, R. (2010). Egg incubation temperature affects male reproductive success but not display behaviors in lizards. *Behavioral Ecology and Sociobiology*, *64*, 803-813.

Wierzbicka, A. (1996). *Semantics: Primes and universals*. Oxford University Press, New York

Wiley, R. H. & Richard, D. G. (1978). Physical constraints on acoustic communication in the atmosphere: Implications for the evolution of animal communication. *Behavioral Ecology and Sociobiology*, *3*, 69-94.

Willigen, R. F., van der Frost, B. J. & Wagner, H. (2003). How owls structure visual information. *Animal Cognition*, *3*, 39-55.

Wong, B. B. & Rosenthal, G. G. (2006). Female disdain for swords in a swordtail fish. *The American Naturalist*, *167*, 136-140.

Woo, K. L., Burke, D. & Peters, R. A. (2009a). Motion sensitivity in lizards: Random-dot kinematograms reveal the importance of motion noise for signal detection. *Animal Behaviour*, *77*, 307-315.

Woo, K. L., Hunt, M., Harper, D., Nelson, N. J., Daugherty, C. H. & Bell, B. D. (2009b). Discrimination of flicker frequency rates in the reptile tuatara (*Sphenodon*). *Naturwissenschaften*, *96*, 415-419.

Woo, K. L. & Rieucau, G. (2008). Considerations in video playback design: Using optic flow analysis to examine motion characteristics of live and computer-generated animation sequences. *Behavioural Processes*, *78*, 455-463.

Woo, K. L. & Rieucau, G. (2011). From dummies to computer-animated stimuli: A review of computer-animated stimuli used in animal behavior studies. *Behavioral Ecology & Sociobiology*, *65*, 1671-1685.

Woo, K. L. & Rieucau, G. (2012). Aggressive signal design in the Jacky dragon (*Amphibolurus muricatus*): Display duration affects efficiency. *Ethology*, *118*, 157-168.

Woo, K. L. & Rieucau, G. (2013). Efficiency of aggressive and submissive visual displays against environmental motion noise in the Jacky dragon (*Amphibolurus muricatus*). *Ethology, Ecology & Evolution*, *25*, 82-94.

Woo, K. L. & Rieucau, G. (2014). The importance of syntax in a dynamic visual signal: Recognition of Jacky dragon threat displays depends upon sequence. *Acta Ethologica*. DOI: 10.1007/s10211-014-0209-1

Wood, W. E. & Yezerinac, S. M. (2006). Song sparrow (*Melospiza melodia*) song varies with urban noise. *Auk*, *123*, 650-659.

Yasukawa, K. (1989). Costs and benefits of a vocal signal: The nest-associated "Chit" of the female red-wingned blackbird, *Agelaius phoeniceus*. *Animal Behaviour*, *38*, 866-874.

Zanker, J. M. (1999). Perceptual learning in primary and secondary motion vision. *Vision Research*, *39*, 1293-1304.

Zeil, J. (2000). Depth cues, behavioural context, and natural illumination: Some potential limitations of video playback techniques. *Acta Ethologica*, *3*, 39-48.

Zeil, J. & Zanker, J. M. (1997). A glimpse into crabworld. *Vision Research*, *37*, 3417-3426.

In: Animal Communication and Cognition
Editor: Tabitha Wagner
ISBN: 978-1-63482-411-8

Chapter 2

PRETEND PLAY IN SIGNING CHIMPANZEES (*PAN TROGLOGYTES*)*

Tennyson E. Egan and Mary Lee A. Jensvold†
Central Washington University, Ellensburg, WA, USA

ABSTRACT

Pretend play is well studied in human children and is thought to be associated with symbolic thought and theory of mind. There are many descriptions of pretend play in chimpanzees and other apes but few systematic studies. This study was a systematic sampling of videotapes for instances of pretend play in chimpanzees who have acquired American Sign Language. There were two sampling methods. The first sample of 17 hours 36 minutes and 26 seconds of videotape contained five instances of pretend play. The second sample of 50 hours 16 minutes and 41 seconds contained 16 instances. The instances fell into four different categories; substitution, attribution of function, animation, and insubstantial situation attribution.

---

* Preparation of this manuscript was partially supported by the Office of the Dean, College of the Sciences, Central Washington University, Ellensburg, WA. This research was in partial completion of a Master of Science degree.

† Correspondence concerning this article should be addressed to Mary Lee Jensvold, Department of Anthropology & Museum Studies, Central Washington University, 400 E. University Way, Ellensburg, WA 98926-7573 or maryleejensvold@yahoo.com.

DEFINING PRETEND PLAY

Over the years a great deal of interest in the behavior of play has culminated in numerous studies exploring the topic (Fagen, 1981; Fein, 1975, 1981; Gomendio, 1988; Iwaniuk, Pellis, & Nelson, 2001; Jensvold & Fouts, 1993; Lowe, 1975; Matthews, 1977; Mitchell, 1990, 2002; Nielson & Dissanayake, 2000; Pellegrini & Bjorklund, 2004; Pellegrini & Smith, 1998; Piaget, 1962). One form of play behavior is pretend play. A variety of terms are used interchangeably to describe this type of play behavior: make-believe play, fantasy play, dramatic play, imaginative play, representational play, simulative play, and symbolic play (Fein, 1981; Lowe, 1975; Mitchell, 1990; Piaget, 1962). Pretend play is an intriguing and fairly common behavior observed in human children.

Many researchers have described and defined pretend play (Lillard, 1993; McCune-Nicolich, 1981; Mitchell, 2002; Pellegrini & Bjorklund, 2004; Pellegrini & Smith, 1998). Lillard (1993) described pretend play as the intersection between play and pretense. She offered a useful definition of pretend play: "the projecting of a supposed situation onto an actual one, in the spirit of fun rather than for survival" (p. 349). Pellegrini and Smith (1998) recommended a more detailed definition of pretend play: "Fantasy involves an 'as if . . .' orientation to the world and involves actions, use of objects, and verbalizations with nonliteral meanings; often, it involves playing a distinct pretend role such as mummy, fireman, doctor" (p. 52). McCune-Nicolich (1981) offered a checklist of specific criteria commonly used to infer when a child is engaged in pretend play:

> (1) inanimate objects are treated as animate (e.g., caretaking of a doll), (2) everyday activities are performed in the absence of the necessary materials (e.g., drink from an empty cup), (3) the child performs actions usually done by someone else (e.g., cooking, telephoning), (4) activities are not carried to their usual outcome (e.g., purse over arm, wave, but not go out), (5) one object is substituted for another (e.g., shell = cat), and (6) affective and instrumental behaviors by the child signal the nonliteral quality of the activity. (pp. 785-786)

McCune-Nicolich further described pretend play as a transformation of "activities from their real objectives and objects from their real counterparts" (p. 786). Matthews (1977) developed six categories of pretend play. These provided objective definitions that can be linked to observed behaviors and were used in this study.

PRETEND PLAY IN COGNITION

Lunzer (1959), McCune-Nicolich (1981) and Piaget (1962) described both pretend play and language as manifestations of symbolic thought since both involve active representations. Smith (1982) argued that the relationship between pretend play and language may actually be a causal relationship. Although there is a positive correlation between development of pretend play and development of language; children without language and children with language-impairments engage in pretend play (Casby & Ruder, 1983; Gregory, 1976; Terrell, Schwartz, Prelock, & Messick, 1984; Udwin & Yule, 1983). Recent theorists suggest pretend play is evidence of theory of mind since the child must take the perspective of another (Pellegrini & Bjorklund, 2004; Smith 2002, 2005; Veneziano, 2002).

There are many theories on the possible function(s) of pretend play (McGhee, 1979; Parker & Gibson, 1979; Pellegrini & Bjorklund, 2004; Piaget, 1962; Smith, 1982). Pellegrini and Bjorklund (2004) suggested that pretend play has an "immediate function of helping juveniles take the perspective of other players" (p. 29). Smith (1982) theorized that pretend activity "optimized the degree of complexity and goal-directedness of play behavior as functional practice for adult skills" (p. 152). Along the same lines, Parker and Gibson (1979) argued that pretend play is vital for the rehearsal of subsistence roles. Piaget's (1962) observations that during pretend play young children often enact daily activities such as eating, preparing and serving food, and sleeping, support this argument. McGhee (1979) asserted a somewhat different function when he noted that in pretend play children could create a new and interesting set of circumstances by manipulating or rearranging some aspect of reality. Similarly, Piaget (1962) speculated that pretend play allows children to reproduce a modified version of reality for pleasure. It is doubtful that pretend play has just one specific function; more likely several exist (Bekoff, 1998; Fagen, 1981).

PRETEND PLAY IN NON-HUMAN PRIMATES

The existence of pretend play in non-human primates contributes to our understanding of this behavior in humans ontogenetically and phylogenetically. There are many descriptions of pretend play in chimpanzees (R. Gardner & Gardner, 1969, 1989; Goodall, 1986; Hayaki, 1985; Hayes,

1951; Sandel, 2010; Wrangham, 1995), bonobos (Lyn, Greenfield, & Savage-Rumbaugh, 2006), gorillas (Matevia, Patterson, & Hillix, 2002), and orangutans (Russon, 2002). Please see Gomez and Martin-Andrade (2005) for review.

While descriptions of pretend play are numerous, there are few systematic studies. Jensvold and Fouts (1993) studied pretend play in 5 signing chimpanzees. The researchers recorded the chimpanzees' behavior with a remote videotaping technique. They put a variety of objects in the enclosure each day to simulate a playroom environment. Fifteen hours of behavior were recorded over a 3-week period. There were six instances of pretend play as defined by Matthews' (1977) and were classified into two categories: animation and substitution. The definitions used in Matthews' study are described in section 5.1.3 and Table 3.

Mignault (1985) compared the behavior of nursery-reared chimpanzees to that of human children when interacting with objects typical of their environment. The researcher watched videotapes of the chimpanzees' behavior and classified the behaviors into one of three categories: activities without objects, interpretable activities with objects (the functional use of an object and "make-believe" play) and non-interpretable activities with objects (activities in which the subject used an object on him or herself, did not use an object on him or herself, and used more than one object). Interestingly, Mignault did not record the behaviors of brushing a doll's hair and body and acting as if eating from a plate with a spoon handle as pretend play. She concluded that the chimpanzees did not exhibit pretend activity.

Lyn, Greenfield, & Savage-Rumbaugh (2006) examined the development of pretend play in chimpanzees and bonobos using archival videotape. They found 27 examples in both species representing all the developmental stages in the scheme. The purpose of this study was to build upon Jensvold and Fouts' (1993) pretend play research. While Jensvold and Fouts analyzed 15 hours of videotape recorded over a 3-week period, this study was a longitudinal analysis of more than 67 hr of videotape recorded over an 18-year period.

General Method

Overview

There were two studies to sample videotapes of chimpanzee behaviors for instances of pretend play. The participants were the same individuals in each study. An observer viewed videotapes for instances of pretend play and then classified those instances in both studies. The procedure to select and sample videotapes differed between the two studies. Study 2 ensured an equal sampling of videotapes between the years of the study. Both studies resulted in a descriptive analysis of the instances.

Participants

The participants in this study were five chimpanzees (*Pan troglodytes*) who lived at the Chimpanzee and Human Communication Institute (CHCI) located at Central Washington University (CWU) in Ellensburg, WA. This community of chimpanzees comprised three females and two males. Table 1 contains biographical information for each chimpanzee. The chimpanzees had lived together at CWU since 1981.

Table 1. Biographical Information for Chimpanzee Participants

	Washoe	Moja	Tatu	Dar	Loulis
Birth date	10/65 (est.)	11/18/72	12/30/75	8/2/76	5/5/78
U. of Nevada-Reno[a]	1966-70	1972-79	1975-81	1976-81	n/a
U. of Oklahoma[b]	1970-1980	1979-1980	n/a	n/a	1979-1980
Central Wash. U[c]	1980-2007	1980-2002	1981-2013[d]	1981-2012	1980-2013[d]

[a] Cross-fostered, exposed to ASL only.
[b] Exposed to ASL and spoken English.
[c] Exposed to ASL and spoken English until 6/86; thereafter exposed primarily to ASL.
[d] Relocated to Fauna Foundation, Montreal, Quebec June, 2013.

Four of the five chimpanzees observed in this research, Washoe, Moja, Tatu and Dar, were raised like human children using an ethological technique called cross-fostering (R. Gardner & Gardner, 1989). The Gardners described this approach as follows:

> Cross-fostering a chimpanzee is very different from keeping one in a home as a pet. Many people keep pets in their homes. They may treat their pets very well, and they may love them dearly, but they do not treat them like children. True cross-fostering – treating the chimpanzee infant like a human child in all respects, in all living arrangements, 24 hours a day every day of the year – requires a rigorous experimental regime that has rarely been attempted. (R. Gardner & Gardner, 1998, p. 292)

The cross-fosterlings drank from cups, ate at the table with silverware, dressed themselves, used a toilet and had the usual children's toys (R. Gardner & Gardner, 1989).

In addition to being treated like children, the cross-fosterlings were immersed in American Sign Language (ASL). The cross-fostered chimpanzees learned the signs of ASL through modeling, much the same way human parents model speech and sign for human children. Human caregivers signed to each other and to the cross-fosterlings throughout the day and commented frequently on common objects and events in the environment (R. Gardner & Gardner, 1989).

As a young adult Washoe adopted 10-month-old Loulis, the fifth chimpanzee in this research. To determine whether Loulis would acquire signs without human intervention, all human signing, except for seven signs, WHO, WHAT, WHERE, WHICH, WANT, SIGN, and NAME, was prohibited in his presence. Loulis spent all of his time with Washoe and other signing chimpanzees. He began to sign in seven days and combined signs into phrases in five months. In the 5-year-period of signing restriction, Loulis learned 51 signs (Fouts, Hirsch, & Fouts, 1982; Fouts, Fouts, & Van Cantfort, 1989). Like the cross-fostered chimpanzees and human children, Loulis acquired his signs in a conversational setting and later he used his signs in conversations with human caregivers and with the other chimpanzees (Fouts & Fouts, 1989; Leitten, Jensvold, Fouts, & Wallin, 2010; Leeds & Jensvold, 2013; Jensvold, Wilding, & Schulze, 2014).

As adults the cross-fostered chimpanzees lived on the campus of CWU in Ellensburg. They continued to sign with caregivers in all interactions (Bodamer & Gardner, 2002; Jensvold & Gardner, 2000; Leitten et al. 2010), to each other (Fouts & Fouts, 1989; Jensvold et al., 2014; Leeds & Jensvold, 2013), and to themselves (Bodamer, Fouts, Fouts, & Jensvold, 1994). Activities in their daily lives included meals, games, and parties. At the time of this writing, the surviving chimpanzees, Tatu and Loulis, live at the Fauna Foundation in Carignan, Quebec, Canada.

Facilities

From January 1, 1985 to May 6, 1993, the chimpanzees lived on the third floor of the Psychology Building on the campus of CWU. The chimpanzee area in this facility was 27.87 m^2 and contained tunnels and enclosures. The chimpanzees had access to all of the enclosures at all times. From May 7, 1993 through the end of the study period the chimpanzees lived in CHCI building on the campus of CWU. The chimpanzee area at CHCI measured 587 m^2, with 4,417 m^2 of climbable surface area. It comprised four inter-connecting night enclosures, two large indoor areas, one large outdoor area, and human interaction areas.

The indoor and outdoor areas contain climbing structures, cargo nets, tires, and platforms thus creating a complex environment. The chimpanzees had free access to these areas throughout the day. However, they were invited into their night enclosures for meals and nesting at night and did not normally have access to the other areas during this time.

The chimpanzees in this study, like children in pretend play research (Fein, 1975; Lowe, 1975; Matthews, 1977; Nielson & Dissanayake, 2000), had access to a variety of play materials each day. These included toys, stuffed animals, magazines, clothes, mirrors, containers, hoses, bags, purses, brushes, and coloring materials. Given that the videotapes were recorded over an 18-year period, there was a wide variety of play materials that the chimpanzees utilized. Most of the play materials were familiar to the chimpanzees; however, they also had novel items.

Video Database

The chimpanzees' caregivers recorded the database of videotapes used in this study. This video recording was part of the regular continuous record keeping. Camera operators recorded interesting chimpanzee behaviors and interactions, such as signing, tool use, play, fights, and grooming and all special events, such as holidays and birthday parties. The amount of recording on each videotape varied; some videotapes contained a full 2 hr and others had 2 minutes.

Also, the amount of time that each chimpanzee appeared on each videotape varied. The database included 245 videotapes recorded between 1985-2002 (4 from 1985, 18 from 1986, 16 from 1987, 6 from 1988, 5 from 1989, 13 from 1990, 5 from 1991, 12 from 1992, 40 from 1993, 13 from 1994, 21 from 1995, 13 from 1996, 31 from 1997, 12 from 1998, 11 from 1999, 17 from 2000, 4 from 2001, and 4 from 2002).

STUDY 1

Method

Videotape Selection

An observer (T.E.), with 2 years of training and experience in chimpanzee behaviors and ASL, viewed videotapes for the presence of pretend play. In the first phase of study 1, T.E. viewed eight videotapes for pretend play. To randomly select videotapes, she assigned to each videotape a number between 1 and 245. She then selected eight videotapes with numbers that matched the first eight numbers on a random number list. T.E. then viewed the entire eight videotapes. This procedure meant that years of the study were randomly, but not equally, represented.

Table 2. Videotape Group Description and Number of Instances in Studies 1 and 2

Group	1		2		3		4	
Years	1985-1988		1989-5/6/1993		5/7/1993-1997		1998-6/2/2002	
Tape no.	1-44		45-87		88-197		198-245	
Individual	S1	S2	S1	S2	S1	S2	S1	S2
Washoe	0	0	0	0	0	0	0	1
Moja	0	1	0	0	1	0	2	2
Tatu	0	1	0	2	1	2	0	1
Dar	0	0	0	0	0	1	0	0
Loulis	1	4	0	1	0	0	0	0
Total	1	6	0	3	2	3	2	4

Note. S1 is study 1 and S2 is study 2.

In the second phase of study 1 a different procedure of selection was used that ensured a more equal sampling of years. The database of videotapes was divided into four groups according to the year the videotapes were recorded each group representing roughly four years. The years each group included were: group 1 1985-1988; group 2 1989-May 6, 1993; group 3 May 7, 1993-1997; and group 4 1998-June 2, 2002. Table 2 contains information on each group of videotapes. Each videotape was assigned a unique number between 1-245. T.E. selected a videotape when its corresponding number appeared on a list of random numbers. Once a videotape from a group was selected, only videotapes from the remaining unselected groups were eligible for selection. She repeated this process until a videotape from each group was selected.

Then she repeated the selection process. This selection process ensured an equal and random sampling from each group of videotapes. T.E. viewed only the first 15 min of each selected videotape. Seven of the videotapes were less than 15 min in length; therefore, they were viewed in their entirety. T.E. viewed 23 videotapes from each of the four groups using this method. She then calculated the total amount of viewing time for each group.

Locating Instances of Pretend Play

T.E. used six categories of pretend play as an operational definition of pretend play (Matthews, 1977) and they are described in Table 3. When T.E. observed an instance of pretend play she recorded the time on the videotape that it appeared and the name of the acting chimpanzee. She then made a detailed description of the behavior. The CHCI Behavioral Taxonomy (McCarthy, Jensvold, & Fouts, 2012) contains descriptions of chimpanzee behaviors and T. E. used this to describe the behaviors that she observed. She then recorded any sign glosses that occurred during the instance and notes on sign modulation. The gloss is the English word equivalent for a sign. She used the place, configuration, and movement (PCM) system (B. Gardner, Gardner, & Nichols, 1989) to identify sign glosses. B. Gardner and Gardner (1994) and Rimpau, Gardner, and Gardner (1986) described how cross-fosterlings inflected signs and phrases which included reiteration and changes in sign placement and T.E. noted sign modulation on reiterations or place changes. Transcriptions of signs in this article are written in capital letters followed by notes on modulation in parenthesis. A gloss followed by "x" indicates immediate reiteration of that sign.

Categorizing Instances of Pretend Play

Finally T.E. categorized the instance of pretend play. Table 4 contains descriptions and examples of Matthews' six categories of pretend play. Matthews' first category was substitution defined as "the attribution of an entirely new identity to a referent" (p. 214). Matthews' second category was attribution of function defined as "the ascription of a functional property to a referent that does not actually possess that property" (p. 214). Matthews' third category was animation defined as "the attribution of human or living characteristics or functions to an inanimate object" (p. 214). These three categories all incorporate materials that actually exist in the playroom situation.

Table 3. Matthews' Six Predominant Categories of Pretend Play

Categories	Definition	Type	Example
Substitution	The attribution of an entirely new identity to a referent.	Material	A child placing a bag on his hand and calling it a glove.
Attribution of function	The ascription of a functional property to a referent that does not actually possess that property.	Material	A child using a toy stove and asking his or her parent, "Do you want your soup heated?"
Animation	The attribution of human or living characteristics or functions to an inanimate object.	Material	A child feeding her toy dog.
Insubstantial material attribution	Reference to materials that do not actually exist, at least not in the present playroom situation.	Ideational	A child exclaiming that magic was clogged in a toy horn.
Insubstantial situation attribution	Reference to situational factors not actually existing in the context of the playroom and play session. Also includes singing and rhyming for the present study.	Ideational	A child pounding on a cable spool with a hammer and announcing fireworks were going to occur. Or a child saying "A cat cat cat ate a rat rat rat."
Character attribution	Portrayal of the qualities of a character by active representation.	Ideational	A child acting as if he or she were a police officer.

Table 4. Number of Pretend Play Instances In Each Category for Studies 1 and 2

	Substitution		Attribution of Function		Animation		Insubstantial Material		Insubstantial Situation		Character Attribution		Total	
Individual	S1	S2	S1	S2	S1	S2	S1	S2	S1	S2	S1	S2	S1	S2
Washoe	0	0	0	0	0	0	0	0	0	1	0	0	0	1
Moja	2	0	1	1	0	1	0	0	0	1	0	0	3	3
Tatu	0	0	0	0	0	1	0	0	1	5	0	0	1	6
Dar	0	0	0	0	0	1	0	0	0	0	0	0	0	1
Loulis	0	0	0	0	1	5	0	0	0	0	0	0	1	5
Total	2	0	1	1	1	8	0	0	1	7	0	0	5	16

Note. S1 is study 1 and S2 is study 2.

The remaining three categories all share the characteristic of ideation and do not necessarily incorporate present materials. As Matthews' described "the actual referent of the fantasy event is intangible...and exists only in the virtue of the child's imaginative act" (p. 214). Matthews' fourth category was insubstantial material attribution defined as "reference to materials that do not actually exist, at least not in the present playroom situation" (p. 214). Matthews' example was a child sitting on the floor, tapping the carpet with a toy horn and exclaiming that the magic was clogged in the horn. In this case, the material that the child referred to, magic, did not exist. Matthews' fifth category was insubstantial situation attribution defined as "reference to situational factors not actually existing in the context of the playroom and play session" (p. 214). For this study, this category also included the reiteration of signs and sign rhyming, as was included in Jensvold and Fouts' (1993) research on pretend play in this same group. As they described:

> Bodamer, et al.'s (1994) study of private signing in these chimpanzees provides examples of Insubstantial Situation Attribution. For example, Moja signed RED RED RED CRY CRY CRY FUNNY to herself as she lay down. Signs of ASL can be broken down into components of the place where the sign occurs, the configuration of the hand, and the movement that occurs (B. Gardner, Gardner & Nichols, 1989; Stokoe, Casterline & Croneberg, 1965). When some aspects of these components are similar between signs it is analogous to spoken rhyming (Klima & Bellugi, 1979). The three signs in Moja's above instance of imagination rhyme in that the handshape (index finger extended from fist) and the movement (the tip of the index finger rubs down the place) are the same. The only difference between these signs is the place where they occur. Thus this instance is analogous to the alliterations often found in poems. The repetition of the signs also contributes to the instance's songlike quality as children are often observed repeating words to themselves in their songs. p. 223

Matthews' sixth category was character attribution defined as "portrayal of the qualities of a character by active representation" (p. 214).

Results

In 17 hr 36 min and 26 sec of videotape, Moja, Tatu, and Loulis together exhibited five instances of pretend play. The instances fit into four of Matthews' categories: substitution, attribution of function, animation, and insubstantial situation attribution. Table 4 shows the frequency of each category for each chimpanzee. The instances appeared in videotape groups 1, 3, and 4. Table 2 shows the number of instances in each group. Because the first phase of study 1 did not equally sample each group, there were no further analyses in this study. A brief description of each instance appears in Table 5.

Table 5. Brief description of instances in each category

Animation
S1 (8804M02) 1:02:42-1:06:22
Loulis was pilo erect and displayed towards a box with pilo erect hair, panthooting. He repeated hits the box with brief pauses to eat lilacs. He also bits and hits the box while donning a playface.
S2 (8604M02) 56:27-57:04
Loulis played with a blanket scooting on it, spreading it, hitting it while wearing a playface.
S2 (8601M04) 10:47-11:55
Moja held a mirror and signed PEEKABOO five times during the session. The sign PEEKABOO covers the eyes. At times she parted her fingers or took her hand away from her face at the end of the sign. The sign PEEKABOO effectually is the game when looking in the mirror.
S2 (8701M05) 1:39:17-1:42:17
Loulis played with a purse. Initially he bit and flailed it. This escalated to displays and hits toward it with a playface.
S2 (9304M08) 1:43:49-1:44:07
Dar picks up a doll and kissed it.
S2 (9301M01) 22:52-24:35
Loulis displayed toward a box and flailed and hit it.
S2 (8504M03) 52:47-53:06
Tatu placed a mask on the floor in front of her. She signed THAT on the mask. She picked up the mask and kissed the mouth of the mask. She then signed THAT four times on the mask.
S2 (8701M06) no time code on Beta tape
Loulis had a shoe. She beat, bit, pressed, bent, and flailed the shoe while wearing a playface.
S2 (8701M06) no time code on Beta tape
Loulis had a purse. He beat and flailed it while wearing a playface.
Attribution of Function
S1 (9804M08) 4:36-5:01
Moja picks up a toy camera holds it to her eye. She pressed the top button with her index finger. She looks at the camera operator. Then she held the camera to her eye.
S2 (9901M01) 0:00-6:23
Moja lay down and held a telephone receiver to her ear while she moved her lips. She

pressed her knuckles on the buttons on the phone base. This continued for the duration. She rearranged the receiver position to her ear and manipulated the phone cord.
Substitution
S1 (9403M01) 0:00-0:38
Moja opened a purse and put it on her head. She looked around, then brushed her arm with a hairbrush while wearing the purse. The hat fell off her head.
S1 (9804M06) 1:52:01-1:55:25
Moja has two linked plastic toy rings (the type used to hang infant toys). She placed one by separating the ends and clipping it on her right earlobe. She shook her head and touched the rings. She removed it and replaced it. Then she looked at herself in a small handheld mirror. She manipulated the ring with one hand while holding the mirror in the other and looked at herself. She then placed the link on her left earlobe and looked in the mirror. She removed the link and unlinked the two rings. She put a single ring on her right earlobe and looked at herself in the mirror. She removed the ring and put it on her left earlobe and looked in the mirror. She removed the ring and walked away.
Insubstantial Situation Attribution
S1 (9304M03) 13:59-15:58
Tatu holds a red hat. She signs THAT with13 reiterations towards the hat. She put the hat on her hand and signed THAT with reiterations. She put the hat on her foot and signed THAT with reiterations and then THAT RED.
S2 (9901M01) 36:01-37:39
Moja wore a dress with a flower pattern. She picked up a mirror and looked at her self throughout this session. She signed THAT (on the dress) FLOWER. She then manipulated the dress. She repeated the utterance THAT FLOWER 10 more times during the session.
S2 (9003M04) 8:14-8:53
Tatu leaned over a mirror that was on the floor and looked at her image throughout this session. She signed THAT on the mirror, then touched her lips. She repeated this and made a playface. She blew her breath on the mirror then signed THAT five times and RED. Then she pulled on one ear then both ears while looking in the mirror.
S2 (9402M02) 0:00-1:36
Tatu had a calendar. She signed THAT on the calendar and THAT BLACK 12 times during the session. The signs within the phrases were reiterated.
S2 (9300M01) 23:00-24:07
Tatu looked at a magazine. She signed THAT (on the magazine). Then THAT BLACK. She repeated these signs throughout the session, at one point reiterating the sign THAT 12 times.
S2 (9204M04) 54:15-55:28
Tatu had a large black plastic hair clip. She clipped it on her right foot and signed THAT BLACK on the clip and reiterated the signs. She looked in a mirror on the floor and reiterated the signs BLACK THAT on the clip. She reiterated these signs. Then she put the clip in her hair.
S2 (0001M01) 14:35-15:50

Table 5. (Continued)

Washoe interacted though glass with a caregiver who was in an observation area. Washoe signed to the caregiver HURRY GIMME. She then "tickled" the caregiver's shoe through the glass. Washoe signed SHOE HURRY two times then HURRY GIMME SHOE and then "tickled" the shoe some more.
S2 (0001M01) 17:03-17:41
Tatu interacted though glass with a caregiver who was in an observation area. The caregiver held a mask up to the glass. Tatu "hit" the mask and then toward the caregiver's face repeatedly. Tatu signed THAT toward the mask. She "hit" the mask again.

STUDY 2

Method

Videotape Selection

As in study 1, T.E. viewed videotapes for the presence of pretend play. The procedure to select videotape was identical to the second phase of study 1. Once a videotape was selected from each group, T.E. viewed the entire tape for pretend play. T.E. viewed 11 videotapes from each of the four groups using this method. She then calculated the total observed minutes for each group. Then T.E. viewed additional videotapes from each group using the same random selection process, until an approximately equal number of minutes were viewed for each group of videotapes.

Reliability

An independent observer (J.K.), with 1 year of training and experience in chimpanzee behaviors and ASL independently coded data to determine interobserver reliability for the presence or absence of pretend play instances. T.E. randomly selected five 2-min segments of videotape with pretend play instances and five 2-min segments without pretend play. For each segment, J.K. indicated the presence or absence of pretend play; reliability was 90%.

The second measure of reliability was for the categorization of pretend play instances. J.K. viewed five segments of videotape containing pretend play and categorized each instance into one of Matthews' six categories. Reliability for the classification of each instance of pretend play was 100%.

Results

In 50 hr 16 min and 41 sec of videotape, all of the chimpanzees together exhibited 16 instances of pretend play. The instances fit into three of Matthews' categories: animation, attribution of function, and insubstantial situation attribution. Table 4 shows the frequency of each category for each chimpanzee. The instances appeared in each videotape group.

Table 2 shows the number of instances in each group. The amount of time analyzed for group 1 was 11 hr 10 min and 25 s, for group 2 was 11 hr and 59 s, for group 3 was 14 hr 16 min and 54 s and for group 4 was 13 hr 48 min and 23 s. The average frequency of pretend play instance per hr was 0.32. The frequency of pretend play instances per hr in group 1 was 0.54, in group 2 was 0.27, in group 3 was 0.21, and in group 4 was 0.29. A brief description of each instance appears in Table 5.

OVERALL DISCUSSION

In studies 1 and 2 the chimpanzees demonstrated pretend play, as defined by Matthews (1977), representing four of the six categories. The categories of pretend play observed in this study were substitution, attribution of function, animation, and insubstantial situation attribution. Neither character attribution nor insubstantial material attribution appeared in this sampling.

Substitution

The two instances of substitution shared the criterion of reidentifying an object. An example of this was on videotape #9804M06.

1:52:01 Moja picked up two interlocked plastic rings with her right hand.
1:52:02 Moja separated the ring ends and placed one ring on her right earlobe.
1:52:11 Moja shook her head with the ring on her ear.
1:52:13 Moja manipulated the ring with her hand while it was on her ear.
1:52:27 Moja shook her head with the ring on her ear.
1:52:29 Moja manipulated the ring while it was on her ear.
1:52:52 Moja removed the ring from her ear.
1:52:55 Moja touched her right ear with her right hand.

1:53:02 Moja picked up the rings with her right hand.
1:53:05 Moja separated the ring ends and placed one ring on her right earlobe.
1:53:12 Moja removed the ring from her ear.
1:53:14 Moja separated the ring ends and placed one ring on her left earlobe.
1:53:20 Moja removed the ring from her ear.
1:53:23 Moja bit the ring.
1:54:21 Moja placed one ring on her right earlobe and looked in a mirror.
1:54:25 Moja manipulated the ring with her left hand while she wore it on her ear and looked in the mirror.
1:54:38 Moja removed the ring from her ear while she looked in the mirror.
1:54:41 Moja separated the ring ends and place one on her left earlobe while she looked in the mirror.
1:54:51 Moja removed the second ring from the first.
1:54:52 Moja placed the first ring on her right earlobe while she looked in the mirror.
1:54:58 Moja removed the ring from her right earlobe and placed it on her left earlobe while she looked in the mirror.
1:55:25 Moja set the ring and the mirror on the floor and quadrupedal walked away from them.

In the other instance of substitution Moja treated a purse as if it were a hat by opening it and wearing it on her head.

Some researchers have argued for an additional criterion in order to demonstrate "true" substitution (Lillard, 1993, p. 350; Mitchell, 2002, p. 9): that the individual knows and understands the conventional use of the object being reidentified. The chimpanzees in this community are aware of the conventional uses for many typical household objects due to the cross-fostering environment they grew up in and subsequent enrichment. They used household tools such as hammers and screwdrivers, and brooms. They observed humans using more "valuable" objects such as cameras and telephones. While the cross-fosterlings wore clothes they did not regularly wear jewelry, makeup, and hats, they did observe their human caregivers accessorize.

Attribution of Function

The two instances of attribution of function shared the criterion of ascribing a functional property to an object that did not actually possess that property. An example of this was on videotape #9901M01 and Figure 1 shows an image from the videotape.

0:00 Moja lay down with a toy telephone held to her ear.
0:03 Moja repeatedly moved her lips.
0:06 Moja pressed her knuckles on the telephone buttons.
0:09 Moja turned the telephone upside down and held it to her ear.
0:16 Moja bit the telephone.
0:52 Moja set the telephone down and pressed and bit the buttons.
1:20 Moja picked the telephone up and held it to her ear.
1:28 Moja pushed the telephone buttons with her thumb.
1:57 Moja held the telephone to her ear while she moved her lips repeatedly.
2:32 Moja repositioned the telephone on her ear.
3:37 Moja moved her lips repeatedly with the telephone to her ear.
3:48 Moja dropped the telephone.
3:50 Moja picked up the telephone and held it upside down to her ear.
4:00 Moja held the telephone right side up to her ear.
4:03 Moja pushed the telephone buttons with her left index finger.
4:19 Moja dropped the telephone.
4:21 Moja held the telephone to her ear.
4:54 Moja manipulated the telephone cord with her left hand.
4:58 Moja moved her lips while she held the telephone to her ear.
5:09 Moja manipulated the base of the telephone with her right hand.
5:14 Moja picked up the base of the telephone with the telephone still to her ear.
5:17 Moja moved her lips repeatedly with the telephone to her ear.
5:33 Moja set the telephone down.
5:37 Moja picked up the base of telephone and bit it.
5:55 Moja set the telephone down and pant hooted.
6:04 Moja picked up the telephone and held it to her ear.
6:23 Moja moved her lips repeatedly with the telephone to her ear.

Figure 1. Moja holds a telephone to her ear.

In the second instance, Moja held a toy camera to her eye while she looked through the viewfinder and pressed the button with her right index finger as if to take a picture. Moja ascribed to the toy camera the functional property of a true camera.

Animation

The nine instances of animation shared the criterion of attributing human or living characteristics to inanimate objects. An example of this was on videotape #9304M08.

1:43:56 Dar picked up a doll from the floor with his left hand.
1:43:58 Dar grabbed the doll with both of his hands.
1:43:59 Dar placed the doll's mouth on his own mouth.
1:44:01 Dar moved his lips in a kissing motion on the doll's mouth with his lips protruded from his mouth.

In the instances of animation Moja, Tatu, Dar and Loulis treated an inanimate object as if it were a living being through such actions as playing and signing peek-a-boo with a mirror, kissing a doll, biting, hitting or flailing shoes and purses while wearing a play face, or displaying mildly towards boxes. A playface is a behavior that clearly and consistently occurs in the social context of play (Goodall, 1986). When it accompanies these actions towards objects it indicates a playful use and suggest and interaction with the objects, indicating the animation of the object.

Animation was the most prevalent category in this study and is well reported in free-living chimpanzees. Wrangham (1995) described a male chimpanzee who played with a log for hours at a time. Wrangham wrote:

> He carried it on his back, on his belly, in his groin, on his shoulders. He took it with him every time he moved. He carried it up four trees, and down again. He lay in his nest and held it above him like a mother with her baby. And he made a special nest that he didn't use himself, except to put the log in. (p. 5)
>
> Hayaki (1985) described a comparable situation in which an adolescent male chimpanzee played with a branch as if he were playing with a conspecific. It included slapping, wrestling, and biting branches while laughing. Sandel (2010) described free-living chimpanzees grooming, cradling, carrying, and copulation behaviors directed to a rock and dead civet.

Gorillas (Matevia, Patterson, & Hillix, 2002) and monkeys also have been reported to demonstrate animation. Breuggeman (1973) observing parental care in a group of free-ranging rhesus monkeys found that many monkeys in this group would press inanimate objects to their chests and "cuddle them." Some of the inanimate objects the monkeys attributed living characteristics to included a styrofoam coffee cup, a small fruit juice can, a crumpled sheet of polyethylene plastic, and a piece of note paper. Koko, a gorilla reared in a highly enriched signing environment, displayed numerous instances of animation (Matevia, et al, 2002).

Insubstantial Situation Attribution

Finally, the eight instances of insubstantial situation attribution shared the criterion of referring to situational factors that were not present in the environment. Two instances of insubstantial situation attribution involved Washoe and Tatu acting like they were tickling a shoe and hitting a mask, respectively. However, neither situation actually occurred because there was glass separating the chimpanzee and the object in which with they interacted. Six of the eight instances incorporated songs, where there was a methodical repetition of one or more signs that gave the signing a songlike quality. An example of this was on videotape #9901M01.

36:01 Moja picked up a mirror and wore a flowered dress.
36:09 Moja manipulated the neck of the dress.
36:12 Moja looked in the mirror and signed THAT (toward the dress) FLOWER.
36:17 Moja held and the mirror with her right hand and manipulated the dress with her left hand.
36:23 Moja looked in the mirror and sign THAT (5x toward the dress) FLOWER.
36:33 Moja looked in the mirror and signed THAT (toward the dress).
36:35 Moja manipulated the dress.
36:39 Moja looked in the mirror signed THAT (toward the dress) FLOWER (2x).
36:46 Moja looked in the mirror and signed THAT (toward the dress) FLOWER.
36:50 Moja looked in the mirror and reiterated THAT (4x toward the dress) FLOWER.
36:55 Moja looked in the mirror.
36:56 Moja looked in the mirror signed THAT (2x toward the dress) FLOWER.
37:01 Moja signed THAT (toward the dress).
37:02 Moja manipulated the dress.
37:08 Moja looked in the mirror and signed THAT (toward the dress).
37:11 Moja smelled the dress.
37:17 Moja manipulated the dress.
37:25 Moja looked in the mirror and signed THAT (2x toward the dress) FLOWER.
37:31 Moja looked in the mirror and signed THAT (toward the dress) FLOWER.
37:39 Tatu climbed on Moja and invited her to play.

The repetition of the utterance and the reiteration of the signs contribute to the singsong quality of Moja's interaction in the mirror. In the other five instances, Tatu reiterated the sign THAT and the utterance THAT BLACK numerous times while she pointed to hair clips and mirrors.

Character Attribution

One category of pretend play that did not appear in this sampling was character attribution. Fein (1981) defined social role enactment as "behavior in which the child simulates the identity or characteristics of another person" (p.

1101). This is equivalent to Matthews (1977) character attribution. Examples of character attribution appear across taxa. For example, a 2-year-old female rhesus followed her mother. The mother carried the infant brother in the ventral position. The daughter clasped a piece of coconut shell with one hand to her ventrum. When her mother stopped and lay on her side while resting one hand on her infant's back, the daughter, just a couple of feet away, adopted the exact posture of her mother, and held the coconut shell to her ventrum (Breuggeman, 1973). The signing gorilla Koko described herself as an elephant adding that a hose was her nose (Matevia, et al, 2002, p. 297). A zoo living gorilla took the role of the human trainer by pretending to feed the trainer the reward with a spoon (Gomez & Martin-Andrade, 2002, p. 265). In addition, there are several cases of social role enactment in Indian Ocean bottlenose dolphins. For example a female bottlenose dolphin frequently imitated the behavior of a fur seal and the postures and swimming behavior of fish, turtles, and penguins (Tayler & Saayman, 1973).

Several authors have described imitating household routines as pretend play which could be classified as character attribution. Part of role playing includes activities that occur in the house (McCune-Nicolich, 1981; Mitchell, 2002; Gomez & Martin-Andrade, 2005). For example Gomez and Martin-Andrade (2005) described: "the partial, nonfunctional imitation of fragments of serious routines, such as cleaning or brushing with the appropriate implements corresponds to some early and simple manifestations of imaginative play in humans . . ." (p. 165). Young Washoe kissed, bathed, and fed her dolls, things that her caregivers did to her (R. Gardner & Gardner, 1969). Viki imitated her adoptive mother by "crudely copying my household routine--dusting, washing dishes, pushing the vacuum cleaner about" (Hayes, 1951, p. 181). In this study, numerous instances were found where the chimpanzees acted "as if" they were cleaning, similar to the daily cleaning routines of their human caregivers but they were not coded as pretend play. In addition, two instances were observed where Washoe blew her breath on a pair of sunglasses and then proceeded to use a piece of fabric to clean the sunglasses, again it was not coded as pretend play. The reason we did not include these was because these cases did not clearly fit into Matthews' (1977) category of character attribution.

Insubstantial Material Attribution

Insubstantial material attribution was the second of Matthews' (1977) pretend play categories that did not appear in this sampling. Yet cases of insubstantial material attribution are reported in other research (Goodall, 1986;

Hayes, 1951). Perhaps the most famous example involves a home-reared chimpanzee named Viki. Hayes (1951) wrote:

> Very slowly and deliberately she was marching around the toilet, trailing the fingertips of one hand on the floor. Now and then she paused, glanced back at the hand, and then resumed her progress. . . . I thought at first that she might merely be enjoying the vibrations from her fingertips . . . But gradually I remembered where I had seen her act this way. Viki was at the pulltoy stage when a child is forever trailing some toy on a string. . . . She trudged along just this busily on two feet and one hand, while the other arm extended backward this way to pull a toy. Viki had an imaginary pulltoy!
>
> No sooner had I arrived at this amazing deduction than she interrupted the sport one day to turn and make a series of tugging motions. That is, they would have been called tugging had there been a rope to tug. . . . Eventually there was a little jerk and off she went again, trailing what to my mind could only be an imaginary pulltoy. (pp. 81-82)

Goodall (1986) described a young free-living chimpanzee:

> Once a four-year-old Wunda watched intently from a safe distance as her mother, using a long stick, fished for fierce driver ants from a branch overhanging the nest. Presently Wunda picked a tiny twig, perched herself on a low branch of a sapling in the same attitude as her mother, and poked her little tool down--into an imaginary nest? (p. 591)

Also Lyn, et al. (2006) described bonobos eating pretend berries (p. 208).

While character attribution and insubstantial material attribution were not found in this study, they have been observed in other groups of apes. The evidence in the systematic reports and the anecdotal cases of pretend play show that chimpanzees engage in the same six categories of pretend play that occurred in Matthews' research with children.

Rates of Occurrence

The frequency of pretend play throughout human development varies. Piaget (1962) argued that pretend activities first emerge during the second year of life, increase during the next several years, and then begin to decline; thus, forming an inverted U-shape pattern. Later research showed the frequency of pretend play relative to other forms of play ranged from 10% to 17% during the preschool years (Nicolich, 1977). During kindergarten years, this frequency increased to approximately 33% (Nicolich) and began to decrease in second grade as compared to the first grade (Eifermann, 1971). In study 2 the rate of pretend play instances decreased from group 1 to group 2 and from group 2 to group 3; however, it increased from group 3 to group 4. All of the chimpanzees in the videotape were well past the age that pretend play peaks in children which may explain the different patterns in this study. Instead this study may have detected the downward slope of the inverted U-shape pattern rather than the entire inverted U-shape pattern.

The overall frequency of pretend play was lower in this community of chimpanzees than in studies of human children. In Matthews' (1977) study, the frequency of pretend play instances per hour per subject was 0.23. In this study, the frequency of pretend play instances per hour per subject was 0.064. In Jensvold and Fouts (1993) the rate for this same group was 0.08. In Lyn, et al. (2006) the rate per hour per subject was 0.05. There is a general consistency among chimpanzees and bonobos and they show a lower rate than children. In Matthews' (1977) study the children were 4-year-olds and the chimpanzees in this study ranged from 6 to 24 years of age, a large difference particularly since pretend play begins to decline at about age 6. A comparison with younger chimpanzees might reveal more similar rates. Also many of the videotapes selected in this study consisted of parties and forages, where the likelihood of pretend play was low due to the overriding theme of celebration. In support of this finding, McCune (1995) found that in children the frequency of pretend play is highly sensitive to contextual elements, and this is likely the case for nonhumans as well. Additionally cultural variations exist in the frequency of pretend play in human children (Pellegrini, Dupuis, & Smith, 2007).

Is Language Necessary?

Numerous examples of pretend play appear in free-living chimpanzees who have no exposure to human language (Goodall, 1986; Hayaki, 1985; Sandel, 2010; Wrangham, 1995) and captive chimpanzees, bonobos, and gorillas who do not use a human induced two way communication (Lyn, et al., 2006; Gomez & Martin-Andrade, 2002). Similarly, children without language and language-impaired children engage in pretend play (Casby & Ruder, 1983; Terrell, et al., 1984; Udwin & Yule, 1983). Yet some have argued that human intervention may be essential for nonhumans to be capable of pretend play (Gomez & Martin-Andrade, 2005; Roberts & Krause, 2002; Smith, 1982, Tomasello & Call, 1997). Perhaps these critics are responding to the large number of records of pretend play in signing and language trained apes as compared to free-living or other captive apes. Veneziano (2002) indicated that in children verbalization during play may make it easier for experimenters to understand or label a behavior as pretend.

> For example in the case of a baby doll placed in a plastic container, if the child chooses to say 'bath' instead of, for example 'baby' or 'here', the child's verbalization makes explicit and brings into focus a particular aspect of her intended pretense activity, an aspect which, to be understood, requires that the other makes the same meaning transformation as the child has imagined. (p. 61).

The same may be true for signing and language trained apes since researchers can use the symbols as a way to understand the situation. In eight of the 21 instances in this study the chimpanzees signed, this is 38% which is considerable. Instances with signs we in all but two of the instances of Insubstantial Situational Attribution. So sign language may increase the frequency of imaginary play, and certainly is essential for aspects of the ISA category. Yet language is not necessary. Additionally, there is more cultural overlap in terms of objects and routines between captive apes and their researchers than between free-living apes and their researchers. With a greater cultural overlap, more behaviors are recognizable to the researchers.

CONCLUSION

This study systematically expanded Jensvold and Fouts' (1993) early observations that the individuals in this community of chimpanzees demonstrate pretend play. Although there are instances of pretend play reported in nonhumans, many instances almost certainly go unnoticed due to either communication barriers (Takhvar & Smith, 1990) or double standards when labeling behavior as pretend play (Mitchell, 1990). The important conclusion in this study is that pretend play, which in children occurs at a low frequency relative to other forms of play (Fein, 1981), is demonstrated by non-human species. This study adds to the growing body of evidence of the occurrence of pretend play in chimpanzees, contrasting claims that pretend play exists only in the repertoire of human primates (Bering, 2001; Fagen, 1981; Rosenberg, 1990; Smith, 1982; Tomasello & Call, 1997).

REFERENCES

Bekoff, M. (1998). Playing with play: What can we learn about cognition, negotiation, and evolution? In D. D. Cummins & C. Allen (Eds.), *The evolution of mind* (pp. 162-182). Oxford, NY: Oxford University Press.

Bering, J. M. (2001). Theistic percepts in other species: Can chimpanzees represent the minds of non-natural agents? *Journal of Cognition and Culture, 1,* 107-137.

Bodamer, M.D., & Gardner, R. A. (2002). How cross-fostered chimpanzees (*Pan troglodytes*) initiate and maintain conversations. *Journal of Comparative Psychology*, 116, 1, 12-26.

Bodamer, M. D., Fouts, D. H., Fouts, R. S., & Jensvold, M. L. A. (1994). Functional analysis of chimpanzee (*Pan troglodytes*) private signing. *Human Evolution, 9,* 281-296.

Breuggeman, J. A. (1973). Parental care in a group of free-ranging rhesus monkeys (*Macaca mulatta*). *Folia Primatologica, 20,* 178-210.

Casby, M. W., & Ruder, K. F. (1983). Symbolic play and early language development in normal and mentally retarded children. *Journal of Speech and Hearing Research, 26,* 404-411.

Eifermann, R. R. (1971). Social play in childhood. In R. E. Herron & B. Sutton-Smith (Eds.), *Child's play* (pp. 270-297). New York: John Wiley.

Fagen, R. (1981). *Animal play behavior*. New York: Oxford University Press.

Fein, G. G. (1975). A transformational analysis of pretending. *Developmental Psychology, 11,* 291-296.

Fein, G. G. (1981). Pretend play in childhood: An integrative review. *Child Development, 52,* 1095-1118.

Fouts, R. (1994). Transmission of human gestural language in a chimpanzee mother-infant relationship. In R. A. Gardner, B. T. Gardner, B. Chiarelli, & F. X. Plooij (Eds.), *The ethological roots of culture* (pp. 257-270). Netherlands: Kluwer Academic.

Fouts, R. (2000). *Chimpanzee & Human Communication Institute Behavioral Taxonomy.* Unpublished manuscript, Chimpanzee & Human Communication Institute, Central Washington University.

Fouts, R. S. & Fouts, D. H. (1989). Loulis in conversation with the cross-fostered chimpanzees. In R. A. Gardner & B. T. Gardner (Eds.), *Teaching sign language to chimpanzees* (pp. 293-307). Albany: State University of New York Press.

Fouts, R. S., Fouts, D. H., & Van Cantfort, T. E. (1989). The infant Loulis learns signs from cross-fostered chimpanzees. In R. A. Gardner & B. T.

Gardner (Eds.), *Teaching sign language to chimpanzees* (pp. 280-292). Albany: State University of New York Press.

Fouts, R. S., Hirsch, A. D., & Fouts, D. H. (1982). Cultural transmission of a human language in a chimpanzee mother-infant relationship. In H. E. Fitzgerald, J. A. Mullins, & P. Gage (Eds.), *Child nurturance* (Vol. 3, pp. 159-193). New York: Plenum Press.

Gardner, B.T., & Gardner, R.A. (1994). Cross-fostered chimpanzees: II. Modulation of meaning. In P.G. Heltne & L.A. Marquart (Eds.), *Understanding chimpanzees* (pp. 234-241). Cambridge, MA: Harvard University Press.

Gardner, B. T., Gardner, R. A., & Nichols, S. G. (1989). The shapes and uses of signs in a cross-fostering laboratory. In R. A. Gardner , B. T. Gardner & T. E. Van Cantfort (Eds.), *Teaching sign language to chimpanzees* (pp. 55-180). Albany: State University of New York.

Gardner, R. A., & Gardner, B. T. (1969). Teaching sign language to a chimpanzee. *Science, 165,* 664-672.

Gardner, R. A., & Gardner, B. T. (1989). A cross-fostering laboratory. In R. A. Gardner & B. T. Gardner (Eds.), *Teaching sign language to chimpanzees* (pp. 1-28). Albany: State University of New York Press.

Gardner, R. A., & Gardner, B. T. (1998). Ethological study of early language. *Human Evolution, 13,* 189-207.

Gomendio, M. (1988). The development of different types of play in gazelles: Implications for the nature and functions of play. *Animal Behaviour, 36,* 825-836.

Gomez, J., & Martin-Andrade, B. (2005). Fantasy play in apes. In A. D. Pellegrini & P. K. Smith (Eds.), *The nature of play* (pp. 139-172). New York: Guilford.

Goodall, J. (1986). *The chimpanzees of Gombe.* Cambridge, MA: Harvard University Press.

Goodman, M. (1976). Toward a genealogical description of the primates. In M. Goodman & R. E. Tashian (Eds.), *Molecular anthropology* (pp. 321-353). New York: Plenum.

Gregory, S. (1976). The deaf child at play. *The deaf child and his family* (pp. 38-60). New York: John Wiley.

Hayaki, H. (1985). Social play of juvenile and adolescent chimpanzees in the Mahale Mountains National Park, Tanzania. *Primates, 26,* 343-360.

Hayes, C. (1951). *The ape in our house.* New York: Harper and Brothers.

Iwaniuk, A. N., Pellis, S. M., & Nelson, J. E. (2001). Do big-brained animals play more? Comparative analyses of play and relative brain size in mammals. *Journal of Comparative Psychology, 115,* 29-41.

Jensvold, M. L. A., & Fouts, R. S. (1993). Imaginary play in chimpanzees (*Pan troglodytes*). *Human Evolution, 8,* 217-227.

Jensvold, M. L. A., & Gardner, R. A. (2000). Interactive use of sign language by cross-fostered chimpanzees (*Pan troglodytes*). *Journal of Comparative Psychology, 114*, 4, 335-346.

Jensvold, M.L., Wilding, L., Schulze, S.M. (2014). Signs of Communication in Chimpanzees. G. Witzany (Ed.), *Biocommunication of animals* (pp. 7-19). Dordrecht: Springer.

Klima, E. S., & Bellugi, U. (1979). Poetry and song in a language without sound. *The signs of language* (pp. 340-372). Cambridge, MA: Harvard University Press.

Leeds, C. A., & Jensvold, M. L. (2013). The communicative functions of five signing chimpanzees (*Pan troglodytes*). *Pragmatics & Cognition 21*:1, 224-247.

Leitten, L., Jensvold, M.L., Fouts, R.S., & Wallin, J. (2010). Contingency in Requests of Signing Chimpanzees (*Pan troglodytes*). *Interaction Studies, 13,* 147-164.

Lillard, A. S. (1993). Pretend play skills and the child's theory of mind. *Child Development, 64,* 348-371.

Lowe, M. (1975). Trends in the development of representational play in infants from one to three years: An observational study. *Journal of Child Psychology and Psychiatry, 16,* 33-47.

Lunzer, E. A. (1959). Intellectual development in the play of young children. *Educational Review, 11,* 205-217.

Lyn, H., Greenfield, P., & Savage-Rumbaugh, S. (2006). The development of representational play in chimpanzees and bonobos: Evolutionary implications, pretense, and the role of interspecies communication. *Cognitive Development, 21,* 199-213.

Matevia, M.L., Patterson, F.G.P., & Hillix, W.A. (2002). Pretend play in a signing gorilla. In R. W. Mitchell (Ed.), *Pretending and imagination in animals and children* (pp. 285-304). Cambridge, UK: Cambridge University Press.

Matthews, W. S. (1977). Modes of transformation in the initiation of fantasy play. *Developmental Psychology, 13,* 212-216.

McCarthy, M., Jensvold, M.L., & Fouts, D.H. (2012). Use of gesture sequences in captive chimpanzee (*Pan troglodytes*) play. *Animal Cognition, 16(3),* 471-481. doi: 10.1007/s10071-012-0587-6.

McCune, L. (1995). A normative study of representational play at the transition to language. *Developmental Psychology, 31,* 198-206.

McCune-Nicolich, L. (1981). Toward symbolic functioning: Structure of early pretend games and potential parallels with language. *Child Development, 52,* 785-797.

McGhee, P. E. (1979). *Humor: Its origin and development.* San Francisco: W. H. Freeman.

Mignault, C. (1985). Transition between sensorimotor and symbolic activities in nursery-reared chimpanzees (*Pan troglodytes*). *Journal of Human Evolution, 14,* 747-758.

Mitchell, R. W. (1990). A theory of play. In M. Bekoff & D. Jamieson (Eds.), *Interpretation and explanation in the study of animal behavior* (pp. 197-227). Boulder, CO: Westview.

Mitchell, R. W. (Ed.). (2002). *Pretending and imagination in animals and children.* New York: Cambridge University.

Nicolich, L. M. (1977). Beyond sensorimotor intelligence: Assessment of symbolic maturity through analysis of pretend play. *Merrill-Palmer Quarterly, 23,* 89-99.

Nielson, M., & Dissanayake, C. (2000). An investigation of pretend play, mental state terms and false belief understanding: In search of a metarepresentational link. *British Journal of Developmental Psychology, 18,* 609-624.

Parker, S. T., & Gibson, K. R. (1979). A developmental model for the evolution of language and intelligence in early hominids. *The Behavioral and Brain Sciences, 2,* 367-408.

Pellegrini, A. D., & Bjorklund, D. F. (2004). The ontogeny and phylogeny of children's object and fantasy play. *Human Nature, 15,* 23-43.

Pellegrini, A. D., & Smith, P. K. (1998). The development of play during childhood: Forms and possible functions. *Child Psychology and Psychiatry Review, 3,* 51-57.

Pellegrini, A. D., Dupuis, D., & Smith, P. K. (2006). Play in evolution and development. *Developmental Review*, 27, 261-276.

Piaget, J. (1962). *Play, dreams, and imitation in childhood* (C. Gattegno & F. M. Hodgson, Trans.). New York: Norton.

Roberts, W. P., & Krause, M. A. (2002). Pretending culture: Social and cognitive features of pretense in apes and humans. In R. W. Mitchell

(Ed.), *Pretending and imagination in animals and children* (pp. 269-279). New York: Cambridge University Press.

Rosenberg, A. (1990). Is there an evolutionary biology of play? In M. Bekoff & D. Jamieson (Eds.), *Interpretation and explanation in the study of animal behavior* (pp. 197-227). Boulder, CO: Westview.

Sandel, A. A. (2010). Representational play in wild chimpanzees (*Pan troglodytes*) [Abstract]. *American Journal of Physical Anthropology, 50,* 205.

Smith, P. K. (1982). Does play matter? Functional and evolutionary aspects of animal and human play. *The Behavioral and Brain Sciences, 5,* 139-184.

Smith, P. K. (2002). Pretend play, metarepresentation and theory of mind. In R. W. Mitchell (Ed.), *Pretending and imagination in animals and children* (pp. 129-141). New York: Cambridge University.

Smith, P. K. (2005). Social and pretend play in children. In A. D. Pellegrini & P. K. Smith (Eds.), *The nature of play: Great apes and humans* (pp. 173-212). New York: Guildord.

Takhvar, M., & Smith, P. K. (1990). A review and critique of Smilansky's classification scheme and the "nested hierarchy" of play categories. *Journal of Research in Childhood Education, 4,* 112-122.

Tayler, C. K., & Saayman, G. S. (1973). Imitative behaviour by Indian Ocean Bottlenose Dolphins (*Tursiops aduncus*) in captivity. *Behaviour, 44,* 286-298.

Terrell, B. Y., Schwartz, R. G., Prelock, P. A., & Messick, C. K. (1984). Symbolic play in normal and language-impaired children. *Journal of Speech and Hearing Research, 27,* 424-429.

Tomasello, M., & Call, J. (1997). *Primate cognition.* New York: Oxford University Press.

Udwin, O., & Yule, W. (1983). Imaginative play in language disordered children. *British Journal of Disorders of Communication, 18,* 197-205.

Veneziano, E. (2002). Language in pretense during the second year: What it can tell us about "pretending" in pretense and the "know-how" about the mind. In R. W. Mitchell (Ed.), *Pretending and imagination in animals and children* (pp. 59-72). New York: Cambridge University.

Wrangham, R. (1995). Ape cultures and missing links. *Symbols* (Spring), 2-20.

In: Animal Communication and Cognition
Editor: Tabitha Wagner
ISBN: 978-1-63482-411-8

Chapter 3

EVOLUTION OF TRP CHANNELS AS MEDIATORS OF PHEROMONE SIGNALING

***Rory Ostrow*[1,*], *Travers Ruel*[2,*], *Tommy Voigt*[1,*], *Cecil James Saunders*[3] *and Robert Hallock*[1,•]**

[1]Dept. of Neuroscience, Skidmore College, Saratoga Springs, NY, US
[2]Dept. of Chemistry, Skidmore College, Saratoga Springs, NY, US
[3]Department of Otorhinolaryngology: Head and Neck Surgery, Perelman School of Medicine, University of Pennsylvania, Philadelphia, PA, US

ABSTRACT

Members of the transient receptor potential (TRP) channel family are involved in pheromone signaling, in both the vomeronasal and the olfactory systems in many species. Since the light-activated TRP and TRPL channels were discovered in *Drosophila*, TRP channel homologues have been identified in many different sensory systems across taxa. For example TRPC2, TRPM4 and TRPM5 are essential for vomeronasal and olfactory system sensing in many animals. Although these channels serve a very different function from their *Drosophila* homologues, their mechanism of action is similar to most TRP channel members utilized in the mammalian chemical senses--activation of a g-protein coupled receptor triggers opening of the TRP channel through the production of a

• E-mail: rhallock@skidmore.edu.
* These authors contributed equally.

secondary messenger, resulting the influx of cations and the depolarization of the sensory cell. Since, pheromone signaling plays a crucial role in the communication of many diverse non-human animals and TRP channel family members are essential for these systems to function, examining the difference between TRPC2, TRPM4 and TRPM5 may reveal how evolution has shaped these sensory systems. This review describes the array of TRP channels important to pheromone sensing, the evolution processes which shaped them, and their involvement in the transduction of pheromone signals.

INTRODUCTION

Pheromones are chemical signals that trigger a specific behavioral response in another member of the species when received (Ishay, 1973; Leon, 1974; Shorey, 1976). These signals are usually very specific, and indicate factors such as age (Coppée et al., 2011), sex (Choe et al., 2012), and health (Ibáñezet al., 2012). Further, pheromones mediate a wide variety of behaviors, ranging from alarm responses and predatory defenses (Robinson, 1987), to mate selection and even the establishment of social hierarchy (Jones and Hamilton, 1998). Pheromone signaling is found across virtually all species of animals, and although their presence has long been questioned in humans, there is growing evidence that human behavior is affected by pheromone-like olfactory signals (see Saxton and Havlicek, 2010 for a review). Pheromone signaling can act through the primary olfactory system (Lopez et al., 2014), but it chiefly acts through the vomeronasal system (see Zufall, 2005 for a review). In both of these systems, pheromonal signals indirectly activate transient receptor potential (TRP) channels, which causes membrane depolarization, which in turn allows for propagation of these signals to other areas of the brain (Zufall, 2005).

The vomeronasal system is heavily involved in pheromone sensing across many species (Francia et al., 2014). The vomeronasal system consists of the vomeronasal organ (VNO), a tubular structure found on the floor of the nasal cavity, which is lined with specialized sensory cells that resemble olfactory receptor neurons, and the accessory olfactory bulbs (For a full review of anatomy see Silver et al., 2010). After stimulation with pheromones, vomeronasal sensory neurons conduct signals to the accessory olfactory bulb, which then relays these signals to portions of the amygdala and hypothalamus (Francia et al., 2014). Those areas of the brain are responsible for the fight-or-flight response, feeding, mating, emotional responses, and control of the endocrine system, can have a wide range of physiological effects on the pheromone receiver.

TRP channels are expressed by sensory neurons where they are essential in processing the intra-cellular signals triggered by pheromone detection (Francia et al., 2014; Liman and Dulac, 2007). Among these, TRPC2, and TRPM5 are involved in transducing pheromone signals (Kiselyov et al., 2010; Lin et al., 2007). Several other TRP channels have also been discovered across species serving functions that include thermosensation (McKemy et al. 2002; Voets et al. 2004), gustation (Liu and Liman, 2003), irritation (Saunders et al., 2014) and olfaction (Liman et al. 1999; Störtkuhl et al. 1999). With the exception of thermosensation, where TRP channels are precisely tuned to open at certain temperatures, the general mechanism of TRP channels in the chemical senses remains the same. Typically, chemicals activate g-protein coupled receptors (GPCRs) and the subsequent release of the coupled g-proteins activates a phospholipase C dependent cascade that initiates the opening of a TRP channel through which pass cations, particularly calcium (Dulac and Torello, 2003). The conserved mechanism and structure of these channels across a wide range of species indicates divergent evolution from a gene isoform possessed by a common ancestor (Montell, 2001). In the case of pheromone sensing, segregation of the receptor machinery and the cation channels into separate genes allows for selective pressures to shape the specificity of GPCRs with risking deleterious alterations to the signal integrating TRP channel. This system allows for precise activation and the elicitation of a particular behavioral and physiological response in the receiver, ultimately leading to a form of communication that can convey pertinent information between conspecifics that would otherwise be difficult to communicate. In this review, we outline the current understanding of the structure, function, and genealogy of TRP channels that are involved in pheromone signaling across species and speculate on the evolutionary events that shaped the TRP channel family.

THE NATURAL HISTORY OF TRP CHANNELS

TRP channel family members serve a sensory function in a variety of organisms across taxa, indicating a long and eventful evolutionary history. The light-activated TRP channels discovered in *Drosophila,* TRP and TRPL were the first to be characterized (Cosens, Manning, 1969; Niemeyer et al., 1996). These channels become activated following light stimulation and share much of their sequence with voltage-gated calcium channels, although the voltage-sensing domain of the synonymous voltage-gated calcium channel is absent (Niemeyer et al., 1996). Similarly, many of the TRP channel family members found in the genome of the nematode worm *Caenorhabditis elegans* share sequence similarity with the pore region of voltage gated calcium channels (Hobert, 2012). Additionally, TRP channel isoforms are expressed in nearly every vertebrate sensory system (Montell, 2001) and all of them are more closely related in sequence homology to voltage gate calcium channels than any other ion channel family (Yu and Catterall, 2004). That sensory systems across the taxa of the animal kingdom are replete with TRP channel isoforms emphasizes the ancient evolutionary origins of the TRP channel superfamily.

The origin of the TRP channel superfamily is so ancient that TRP homologues have even been identified in the genomes of organisms in separate kingdoms. TRP-like channels are also present in the genome of the yeast *Saccharomyces cerevisiae,* and at least two other fungal species (Denis, Cyert, 2002). Although, fungi have no nervous system to detect the external world, their TRP-like channels are important in the cellular signaling pathway responsible for detecting osmotic shock (Denis, Cyert, 2002). Specifically, after *S. cerevisiae* is exposed to hyperosmotic shock Yvc1p, a TRP channel homologue found on vacuoles, opens resulting in an influx of the intracellular signal calcium (Denis, Cyert, 2002). In this instance, Yvclp is roughly analogous to the ryanodine and inositol-1,4,5-triphosphate (IP_3) receptors, which regulate the release of intracellular calcium stores in animal cells. In contrast to Yvclp in yeast, several mammalian TRP channels also appear to play a role in detecting osmotic stress (Wehner, 2006). The presence of functional TRP channels in organisms from distinct kingdoms suggests that a common ancestor of all animals and fungi had, at minimum, one functional TRP channel isoform.

Since the divergence of the animal and fungal kingdoms, the duplication of TRP channel genes has allowed for the diversification of TRP channel function. Gene families, like TRP channels, are thought to be the result of ancient errors in DNA replication that result in two functional copies of the same gene (Zhang, 2003). The presence of two copies of the same gene allows for normally disastrous mutations to occur without resulting in the death of the organism, as one copy of the gene remains unchanged and can carry on normal function after the second is altered (Conant and Wolfe, 2008). Subsequent duplications of these genes provide biological redundancy and allow for further divergence of gene functions via mutations, ultimately resulting in a large gene super family as is observed with TRP channels (Kondrashov, 2012).

Since, several of the major TRP channel subfamilies (TRPV, TRPC, TRPN, TRPA, TRPP) are represented in both invertebrate as well as vertebrates animals (Christensen and Corey, 2007), the diversification of the TRP channel gene family likely predates the existence of a common ancestor of invertebrate and vertebrates. The expression of TRPC2 in the vomeronasal system of the lampreys, an extant jawless fish, suggests that this channel has been important in pheromone detection during the entire history of the vertebrate linage (Grus and Zhang, 2009). Conversely, a TRPM4 homologue is not found in the lamprey genome (Grus and Zhang, 2009), suggesting that its role in odor and pheromone detection is a more recent development. The presence of TRP channel isoforms in across many taxa, as well as the expression of TRPC2 in the VNO of an extant jawless fish, emphasizes that TRP channels have played an essential role in pheromone detection throughout the vertebrate linage.

TRP Channels and Pheromone Sensing

The three TRP channel isoforms involved in pheromone sensing are TRPC2, expressed in the vomeronasal epithelium, and TRPM5 and M4, expressed mostly in the main olfactory epithelium (MOE) and accessory olfactory bulb (AOB). Research shows that TRPC2 knockout mice show similar deficits in pheromone recognition to those with vomeronasal lesions. Specifically, TRPC2 knockouts showed decreases in male-male aggression, increases in male-male mounting, and changes in maternal behavior (Hasen, Gammie, 2011; Zufall et al., 2005). Research on TRPM5 expressed in the MOE of mice has indicated that it is also involved in pheromone sensing (Lin et al. 2007). In humans, the presence of a vestigial VNO indicates that pheromone sensing must be occurring through some other sensory system, namely the olfactory system. Additionally, many of the genes involved VNO transduction, including TRPC2, are only present in the human genome as non-functioning pseudogenes requiring that any socially relevant odors detect by humans occur through a separate pathway (Miller, 2014). Thus, any report of pheromone sensing in humans is most likely mediated by the TRPM5 pathway or alternatively indicative of a completely novel mechanism for pheromone sensing.

Though TRP channels can serve many different functions, their structure and mechanism remain largely conserved across species and taxa. In particular, pheromone related channels such as TRPC2, TRPM4 and TRPM5 share much of their primary structure with the other TRP subfamilies, and the mechanism of signaling remains largely unchanged. Therefore, differences in function could be attributed to small changes in structure, and differential expression in different aspects of the vomeronasal and/or olfactory systems.

STRUCTURE OF TRP CHANNELS

Structural information on TRPC2, TRPM4, and TRPM5 channels is limited, as it is generally limited for membrane-bound proteins, due to the difficulty of recreating the specific *in vitro* conditions when crystallizing proteins for x-ray electron-density mapping (Seddon 2004). Fortunately, there are several recently solved structures of the TRPV1 channel (See Figure 1) (Cao et al., 2013; Liao et al., 2013; Moiseenkova-Bell et al., 2008), so some inferences can be made regarding the structural and mechanistic details of these channel types, based on the sequences homology of the other channel types.

Some important elements of TRP channel structure appear to be broadly conserved. All observed TRP channel structures appear to form tetramers, four individual subunits that assemble to form a full channel (Cheng et al., 2007; Cvetkov et al., 2011; Moiseenkova-Bell et al., 2008). These tetramers can be composed of the same TRP channel isoform or can be heteromers containing a mix of TRP channels. Most commonly, hetero-tetramers are formed by TRP channel isoforms from the same sub-family (Woo et al., 2014) but recent research has suggested tetramers composed of TRP channel isoforms from different subfamilies might also exist (Du et al., 2014).

All TRP channel family members have six membrane-spanning helices, with a pore-forming region located between the fifth and sixth helix (Cvetkov et al., 2011; Harteneck et al., 2001; Yamaguchi et al., 2001). TRPV1 receptor structures indicate that the pore is gated both by a selectivity filter at the top of the pore, and by a narrow hydrophobic gate at the bottom (Cao et al., 2013; Moiseenkova-Bell et al., 2008). This was also observed in TRPV6, where the selectivity filter is loose (Doyle et al., 1998), and is comprised of side chain carbonyls, not backbone carbonyls of these pore lining amino acids (Cao et al., 2013). This structure accounts for the non-selective conductance of these channels, which is found across all of the TRP subfamilies (Voets et al., 2004).

Collectively, these findings indicate that the transmembrane domains, which dictate the conductance and gating of the channels, are structurally conserved across the different TRP channels.

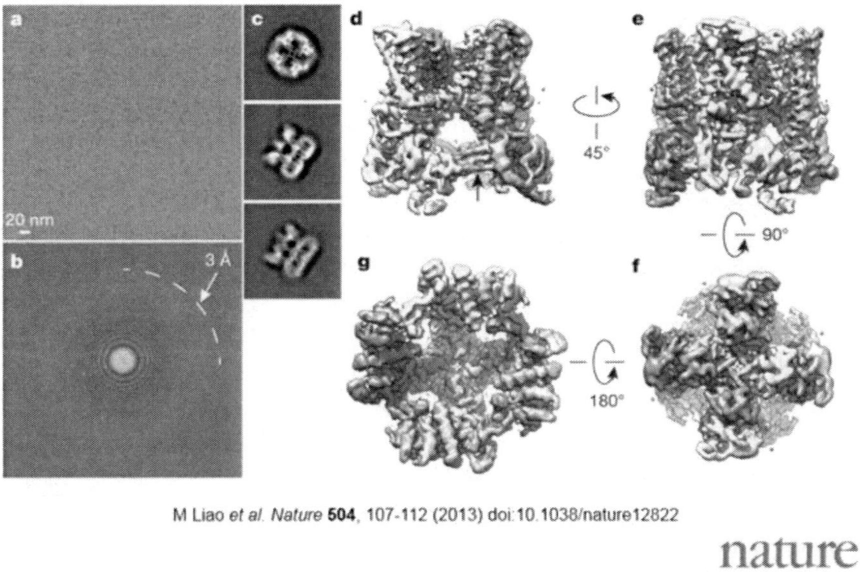

Figure 1. a. Electron micrograph of TRPVI protein. b. Fourier transform of the micrograph in a. c. 2D density maps showing the tetrameric assembly of the subunits. d-g. 3D density maps at 3.4 A with color-coded subunits. Views are from the side (d, e), top (f), and bottom (g).

Although all of TRP channels share a common architecture, there is considerable regional variability between subtypes. TRP channels range from about 750 to 1800 residues in length, with the pore forming transmembrane domains demonstrating the highest degree of conservation between isoforms (Montell 2001; Runnels et al., 2001). In contrast to the ion pore, the intracellular C- and N-terminal domains of TRP channels possess most of the variability found between family members (Runnels 2001).

The high variability in the C- and N-terminal domains imparts the diverse functionality that typifies the members of TRP channel superfamily and allows for members of even the same TRP channel subfamily to have vastly different functions. TRPM6 and TRPM7 are tethered to a fully functional kinase protein in this region (Clémençon 2014). In contrast, the terminal domains of TRPA1 and TRPV channels form a small basketlike structure containing coils and alkyrin repeats (Chang et al., 2004; Jin et al., 2006; Yoshida et al., 2006).

While, the cysteine (thiol containing) residues in TRPV and TRPC channels allow for disulfide bridge formation with thiol-reactive compounds that may contribute to channel gating (Yoshida et al., 2006). These areas may also be involved with coordination of subunit assembly or interactions with other cofactors and secondary messengers (Chang et al., 2004). While this variability makes the TRP channel super family one of the most interesting ion channel families to study, the diversity of functions also makes it challenging to find residues responsible for conferring specific function to each channel type, since there is so little conservation in this region.

Although there is no single residue or region that gives each TRP subtype its function, there are certain small regions of the intracellular domain that are more important than others in determining channel function. For example, single point mutations in the sixth alkyrin repeat in TRPA1 receptors change the channel from cold-activated to heat activated (Jabba et al., 2014). In TRPM channels, there are post-translational modifications such as glycosylation in these regions stabilize the protein (Syam et al., 2014; Woo et al., 2014). Cao et al. (2013) found that TRPV1 receptors form interactions between the C-terminal of the protein and the lipid membrane, and they hypothesized that some agonists might function by breaking this interaction and inducing conformational differences that open the channel. While the intracellular domain varies greatly between channel types, binding sites can usually be found near the C-terminal are important in the agonist specificity found in pheromone-activated TRP channels.

TRPC2 channels are implicated as the primary channel type to promote the activation of neurons in the VNO in response to pheromone binding (Lucas et al., 2003). However, there are no three dimensional structures of TRPC2, or the closely related TRPC1 channels (Kiselyov et al., 2010). This means that certain inferences have to be drawn from other TRP channel structures and TRPC2 sequence analysis to suggest the function of specific TRPC2 domains. For example, TRPC2 is predicted to have an ankyrin-repeat domain near the N-terminus (Miller, 2014). This region is thought to have multiple calmodulin-binding sites (Yildirim et al., 2003), which can act to inhibit channel function (Miller, 2014). There are also several sequence elements that suggest binding sites for diacytlglycerol (DAG), a secondary messenger produced in response to the activation of GPCRs in VNO sensory neurons (Liman et al., 1999).

Similar to TRPC2 in the vomeronasal system, TRPM4 and 5 play a role in odor detection by the olfactory system. In other TRPM channels, subunit assembly is dictated by a coil within the intracellular domain, implying that this region could be similarly important in coordinating assembly in TRPM 4 and 5 (Fujiwara, 2008). Unlike most TRP channels, TRPM4 and M5 channels do not conduct Ca^{2+}, but they are activated by increased concentrations of intracellular Ca^{2+}. This implies the presence of a calcium sensing site in the intracellular domain (Guinamard 2011). TRPM4 is also activated by membrane depolarization, but only in the presence of sufficient Ca^{2+} (Launay et al., 2002). TRPM4 and 5 activation by Ca^{2+} is enhanced by Phosphatidylinositol 4,5-bisphosphate (PIP_2) (Vennekens et al., 2007), which is released by pheromone-activated GPCRs along with DAG (Liu and Liman, 2003).

Mechanism of Activation of TRP Channels

Though TRP channels are the major contributors of transmembrane currents in pheromone sensing cells, TRP channels themselves do not bind pheromones. TRP channels are activated upstream by GPCRs (Lucas et al., 2003). The activation of these GPCRs in the VNO cause an intracellular cascade of events that eventually leads to the influx of ions via TRP channels (Lucas et al., 2003) (see Figure 2). This mechanism allows for more specific activation by their pheromone agonists.

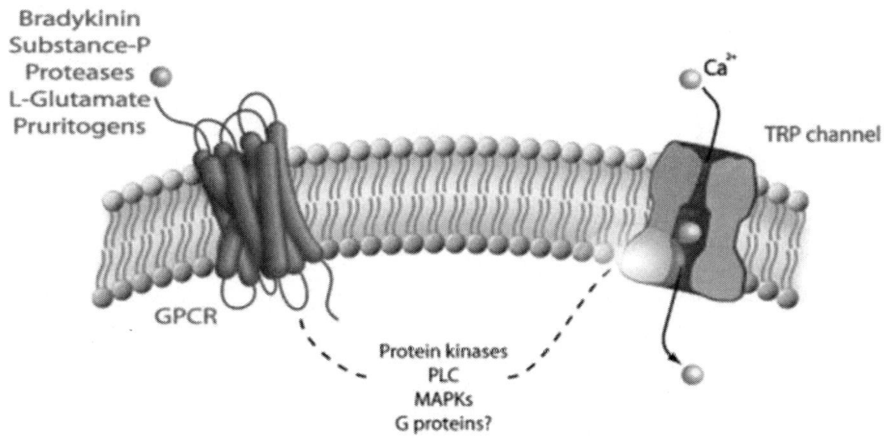

Yekkirala (2013).

Figure 2. Mechanism of modulation of TRP channels through GPCR activation. Binding of agonist to GPCR causes intracellular cascade leading to the influx of Ca^{2+} through TRP channels.

Within the VNO, two major families of pheromone-sensitive GPCRs are expressed by vomeronasal sensory neurons (VSNs) that activate TRP channels: V1R and V2R (Francia et al., 2014). There is much sequence variation in V1R and V2R genes, both between and within species (Ishii, Mombaerts, 2011; Young et al., 2010). Some suggest that this high amount of variation in V1Rs could account for the highly specific nature of binding in the VNO (Ishii, Mombaerts, 2011; Young et al., 2010). Further, slight changes in the structure of the VNO GPCRs can lead to changes in the chemicals that bind these receptors. However, despite these variations in V1R and V2R structures, they still activate the same ion channels, TRPC2. This segregation of function allows for a more precise activation of certain populations of VSNs by certain chemicals, but a more generalized mechanism of depolarization shared by all VSNs. This arrangement increases specificity and allow certain chemicals to have different functions, which is supported by how VSNs communicate with the AOB. Research indicates that different types of VSNs project to different parts of the AOB, making a place map (Jia and Halpern, 1996). This mechanism of coding, and the organization of different TRP channels it affords, leads to further specificity within the VNO.

TRP Channel Expression

Specific TRP channels are expressed in different parts of the pheromonal system and relay specific signals. This is shown by the presence of different TRP channels in different parts of the pheromone transduction pathway. For example, TRPC2 channels are highly concentrated in the VNO while TRPM5 channels are more predominant in the main olfactory bulb (MOB) (Lopez et al., 2014; Zufall, 2005). In addition to TRPC2 in the VNO and TRPM5 in the MOB, TRPM4 is found in the AOB (Shpak et al., 2012). Different parts of the pheromonal pathway will utilize varying subsets of neurons to segregate the pathway controlling differing functions.

Transduction of pheromone-related information occurs in olfactory sensory neurons (OSN) in the MOE. All OSNs have cyclic nucleotide-gated channels (CNG) and among these OSNs, there are either TRPM5 channels or anoctamin 2 channels (ANO2) and OSNs that contain TRPM5 have a function in pheromone signaling (Lopez et al., 2014). Loose-patch current recordings in mouse demonstrated that a TRPM5 inhibitor reversibly and significantly decreased responses to known pheromones (Lopez et al., 2014). Like the MOE, the VNO contains TRP channels involved with pheromone sensing.

The VNO is an important area for pheromone signaling and TRPC2 channels are concentrated in this area (Leypold et al., 2002). TRPC2 deficient mice had absent or diminished VNO field potential responses in response to pheromone mixtures in urine (Leypold et al., 2002). Further, extracellular recordings from vomeronasal sensory neurons in TRPC2 deficient mice failed to show a response to dilute urine (Leypold et al., 2002). Behaviorally, TRPC2 deficient mice mounted other male mice at a much higher rate than wild type mice. There was also decreased aggression in both female and male TRPC2 deficient mice in response to a male intruder. These findings suggest that TRPC2 channels in the VNO have a primary role in regulating sexual and aggressive behavior in mice.

TRP channels can sometimes have more of a subtle role in processing semiochemicals. One example of this is TRPM4 channels in mitral cells of the AOB (Shpak et al., 2012). When stimulated, neurons in the AOB fired for much longer periods than those in the MOB. This prolonged firing was partly due to TRPM4 channels. Specifically, when TRPM4 channels were blocked, the time course in the response of the cells were significantly shortened (See Figure 3) (Shpak et al., 2012). The authors hypothesize that the longer response in the AOB compared to the MOB, was because the MOB analyzed the specific components of a pheromone and that the AOB integrates these components to cause an appropriate behavioral response (Shpak et al., 2012).

TRP channels have also been found to have a role in pheromone detection when there is a varying amount of olfactory stimulus. Oshimoto et al. (2013) showed that mice with a unilateral naris occlusion have a higher number of OSNs that express the TRPM5 channel. Furthermore, in the electro-olfactogram, TRPM5 knockout mice show a decreased pheromonal response in the occluded epithelia compared to wild type mice (Oshimoto et al., 2013). In addition, they showed that CNGA2 knockout mice exhibit a significant increase in the number of TRPM5 channels (Oshimoto et al., 2013). This implies that when there is less input, TRP channels are up-regulated to compensate for the lack of stimulus.

Mitral cells in the MOB that project to the medial amygdala (MeA) in mice react to urine of the opposite sex, which shows their function in pheromone signaling (Thompson et al., 2012). The researchers wanted to see if these neurons are affected by OSNs that express TRPM5. OSNs that express TRPM5 affect the mitral cells in the MOB that project to the MeA. These and other findings show the diversity and expression of TRP channels within the pheromonal system.

Furthermore, these TRP channels specifically relate to pheromone-mediated behaviors. The differential expression of TRP channels throughout the pheromone system can allow for precise activation by a pheromone stimulus, ultimately leading to more precise processing of the stimulus downstream.

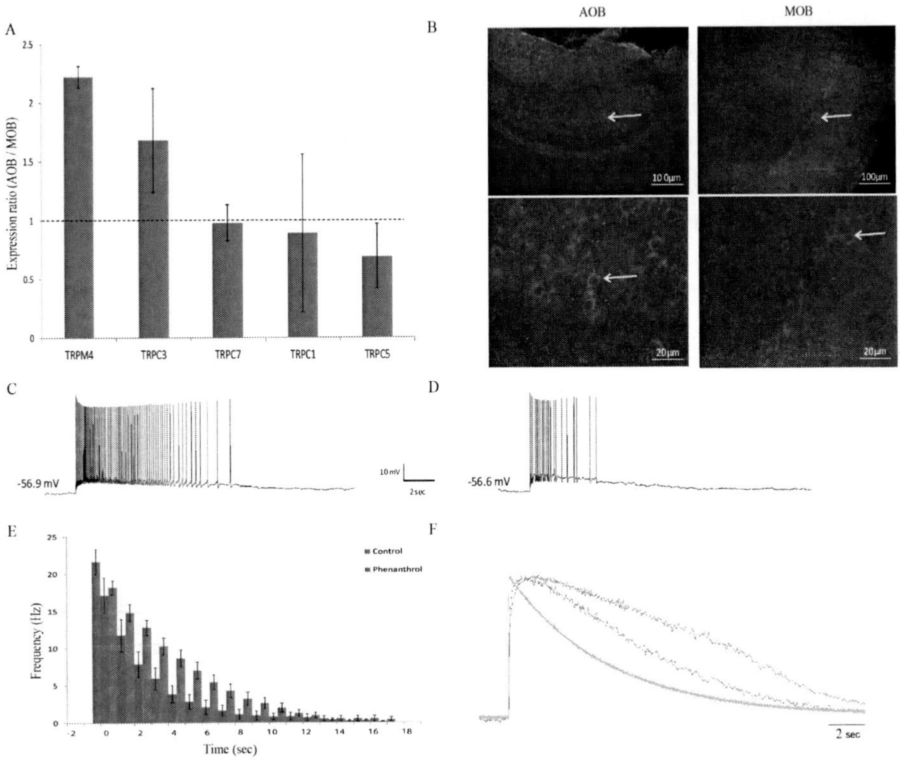

Figure 3. A) Ratio of channel types between the AOB and MOB. B) Immunohistochemical staining of the AOB (left) and MOB (right) for TRPM4 expression. C-D) Voltage responses of an AOB mitral cell to 80 pA currents with the cell unblocked in (C) and blocked in (D). E) Average PSTHs of the responses of eight AOB cells (7 animals) recorded before (blue) and during (red) the presence of a TRPM4 channel blocker. F) Average voltage trajectories of the responses analyzed in (E) in the blocked and unblocked condition.

This allows for animals to be able to recognize pheromones more selectively and leads to more stereotypical behaviors following activation.

DISCUSSION

TRP channels are expressed across essentially every tax in the animal kingdom (Cosens, Manning, 1969; Niemeyer et al., 1996; McKemy et al., 2002), serving functions that include thermosensation (McKemy et al., 2002; Voets et al. 2004) gustation (Liu and Liman, 2003), irritation (Saunders et al., 2014) and olfaction (Liman et al., 1999; Störtkuhl et al., 1999). The structure of TRP channels is fairly conserved across species and taxa, and as such, differences in function are thought to be a result of divergent evolution, following gene duplication events experienced by common ancestors of the modern organisms whose genomes contain TRP channel family members. The differences in function between subtypes of TRP channels can be attributed to differences in several specific regions of TRP channel genes. Even with differences in function, the mechanism of activation of TRP channels remains largely unchanged. Activation of a GPCR causes an intracellular cascade that eventually leads to the activation of TRP channels and the influx of cations (Lucas et al., 2003). The activation of TRP channels by GPCRs allows for pheromone selectivity while maintaining the integrity of the signaling cascade downstream of the pheromone receptors. VSNs express one type of GPCR that selectively binds to pheromones, and this binding causes the more generalized depolarization of the cell via TRP channels. In addition, the differential expression of TRP channels in different parts of the pheromone system adds to the specificity.

Currently, the three dimensional structures of TRPM5, TRPM4 and TRPC2 are not known (Guinamard et al., 2011 Kiselyov et al., 2010). Discovering the three dimensional structure of these channels could lead to precise mechanistic understanding of their activation and modulation by pheromone-activated GPCRs. This understanding would allow for the production of synthetic pheromones which could be utilized to control invasive organisms or pests (Chiszar et al., 2006).

Further research needs to be conducted into the integration of pheromonal information in the brain. Shpak et al. (2012) hypothesized that the sustained, low frequency firing of action potentials in the AOB caused by TRPM4 channels was due to the AOB integrating a variety of chemical signals and eliciting a behavioral response (Shpak et al., 2012). This is in contrast to the MOB, which separates pheromonal blends into their individual components and analyzes them separately (Dulac and Wagner, 2006). More research needs to be done on this to see if their hypothesis of separate signals in the MOB and integrated signals in the AOB is supported.

Understanding the function, location, mechanism and expression of TRP channels in the pheromone detection system has a wide variety of applications. While in humans TRPA1 is primarily expressed in nociceptive somatosensory fibers, in insects it is also expressed in olfactory receptor neurons (Kwon et al. 2010). Insect TRPA1 is activated by citronellal, which can be used as an insect repellent. Therefore, citronellal could also be used as the basis for creating a designer compound that is targeted to these same receptors (Kwon et al. 2010). Mosquitoes use chemical signals as a cue to orient towards and find humans, thus disrupting this pathway could prevent the spread of diseases like malaria (Gillies and Wilkes, 1969). Since TRP channels play an essential role in integrating pheromone response to different pheromones, even across taxa, disrupting these channels' functioning is a good target for disrupting pheromonal signaling. This could help humans devise devices and improve technology to repel any number of disease-carrying animal species.

Research into the functioning of TRP channels could also be valuable in trapping invasive species. Sex pheromones can be used to trap animals (Shorey et al., 1974), and TRP channels could be a target for general disruption or activation of the pheromone system. In summary, TRP channels, which have tremendous breadth of structure and function, have evolved to mediate highly specific pheromone signaling which serves as a crucial means of intraspecies communication in numerous species.

References

Brann, J. H., Dennis, J. C., Morrison, E. E., and Fadool, D. A. (2002). Type-specific inositol 1,4,5-trisphosphate receptor localization in the vomeronasal organ and its interaction with a transient receptor potential channel, TRPC2. *Journal of neurochemistry*, 83(6), 1452-1460.

Cao, E., Cordero-Morales, J. F., Liu, B., Qin, F., and Julius, D. (2013). TRPV1 channels are intrinsically heat sensitive and negatively regulated by phosphoinositide lipids. *Neuron,* 77(4), 667-679.

Cao, E., Liao, M., Cheng, Y., and Julius, D. (2013). TRPV1 structures in distinct conformations reveal activation mechanisms. *Nature,* 504(7478), 113-118.

Chang, Q., Gyftogianni, E., van de Graaf, S. F., Hoefs, S., Weidema, F. A., Bindels, R. J., and Hoenderop, J. G. (2004). Molecular determinants in TRPV5 channel assembly. *Journal of Biological Chemistry*, 279(52), 54304-54311.

Cheng, W., Yang, F., Takanishi, C. L., and Zheng, J. (2007). Thermosensitive TRPV channel subunits coassemble into heteromeric channels with intermediate conductance and gating properties. *The Journal of general physiology,* 129(3), 191-207.

Christensen, A. P. and Corey, D. P. (July 2007). TRP channels in mechanosensation: direct or indirect activation? *Nature Reviews Neuroscience,* 8, 510-521. doi:10.1038/nrn2149

Chiszar, D., Fox, K., Smith, H. M. (1992). Stimulus control of predatory behavior in the brown tree snake (Boiga irregularis). IV. Effect of mammalian blood. *Behav. Neural. Biol.*, 57(2):167-9.

Clémençon, B., Fine, M., Lüscher, B., Baumann, M. U., Surbek, D. V., Abriel, H., and Hediger, M. A. (2014). Expression, purification, and projection structure by single particle electron microscopy of functional human TRPM4 heterologously expressed in Xenopus laevis oocytes. *Protein expression and purification,* 95, 169-176.

Conant, G. C., and K. H. Wolfe. (2008). Turning a hobby into a job: How duplicated genes find new functions. *Nature Reviews Genetics*, 9.12: 938–950.

Cosens, D., and Manning, A. (1969). Abnormal electroretinogram from a drosophila mutant. *Nature,* 224; 285-287.

Cvetkov, T. L., Huynh, K. W., Cohen, M. R., and Moiseenkova-Bell, V. Y. (2011). Molecular architecture and subunit organization of TRPA1 ion channel revealed by electron microscopy. *Journal of Biological Chemistry,* 286(44), 38168-38176.

Denis, V., and Cyert, M. S. (2002). Internal ca(2+) release in yeast is triggered by hypertonic shock and mediated by a TRP channel homologue. *The Journal of Cell Biology,* 156(1), 29-34.

Doyle, D. A., Cabral, J. M., Pfuetzner, R. A., Kuo, A., Gulbis, J. M., Cohen, S. L., ... and MacKinnon, R. (1998). The structure of the potassium channel: molecular basis of K+ conduction and selectivity. *Science,* 280(5360), 69-77.

Du J., Ma X., Shen B., Huang Y., Birnbaumer L., Yao X. (2014)TRPV4, TRPC1, and TRPP2 assemble to form a flow-sensitive heteromeric channel. *FASEB J.,* 28(11):4677-85. doi: 10.1096/fj.14-251652.

Francia, S., Pifferi, S., Menini, A., Tirindelli, R. (2014). Vomeronasal receptors and signal transduction in the vomeronasal organ of mammals. In Neurobiology of Chemical Communication. Editor C Mucignat-Caretta. Chapter 10. CRC Press, Boca Raton (Fl).

Fujiwara, Y., and Minor Jr, D. L. (2008). X-ray crystal structure of a TRPM assembly domain reveals an antiparallel four-stranded coiled-coil. *Journal of molecular biology,* 383(4), 854-870.

Grus, W. E., Zhang, J. (2009). Origin of the genetic components of the vomeronasal system in the common ancestor of all extant vertebrates. *Mol. Biol. Evol.*, 26(2):407-19. doi: 10.1093/molbev/msn262. Epub 2008 Nov 13.

Guinamard, R., Sallé, L., and Simard, C. (2011). The non-selective monovalent cationic channels TRPM4 and TRPM5. In Transient Receptor Potential Channels (pp. 147-171). Springer Netherlands.

Harteneck, C., Plant, T. D., Schultz, G. (2000). From worm to man: three subfamilies of TRP channels, *Trends in Neurosciences*, 23 (4), 159-166.

Hasen, N. S., and Gammie, S. C. (2011). Trpc2-deficient lactating mice exhibit altered brain and behavioral responses to bedding stimuli. *Behavioural Brain Research*, 217(2), 347-353.

Hinman, A., Chuang, H. H., Bautista, D. M., and Julius, D. (2006). TRP channel activation by reversible covalent modification. *Proceedings of the National Academy of Sciences,* 103(51), 19564-19568.

Hobert, O., (December 27, 2012). The neuronal genome of Caenorhabditis elegans WormBook, ed. The C. elegans Research Community, WormBook, doi/10.1895/wormbook.1.7.1, http://www.wormbook.org.

Holakovska, B., Grycova, L., Jirku, M., Sulc, M., Bumba, L., and Teisinger, J. (2012). Calmodulin and S100A1 protein interact with N terminus of TRPM3 channel. *Journal of Biological Chemistry*, 287(20), 16645-16655.

Ishii, T., and Mombaerts, P. (2011). Coordinated coexpression of two vomeronasal receptor V2R genes per neuron in the mouse. *Molecular and Cellular Neuroscience,* 46(2), 397-408.

Jabba, S., Goyal, R., Sosa-Pagán, J. O., Moldenhauer, H., Wu, J., Kalmeta, B., ... and Grandl, J. (2014). Directionality of Temperature Activation in Mouse TRPA1 Ion Channel Can Be Inverted by Single-Point Mutations in Ankyrin Repeat Six. *Neuron*, 82(5), 1017-1031.

Jia, C., Halpern, M. (1996). Subclasses of vomeronasal receptor neurons: differential expression of G proteins (Giα2 and Goα) and segregated projections to the accessory olfactory bulb. *Brain Research*, 719(1), 117-128.

Jin, X., Touhey, J., and Gaudet, R. (2006). Structure of the N-terminal ankyrin repeat domain of the TRPV2 ion channel. *Journal of Biological Chemistry*, 281(35), 25006-25010.

Kiselyov, K., van Rossum, D. B., and Patterson, R. L. (2010). Chapter Eight-TRPC Channels in Pheromone Sensing. *Vitamins and Hormones*, 83, 197-213.

Kondrashov, F. A. (2012) Gene duplication as a mechanism of genomic adaptation to a changing environment. *Proc. Biol. Sci.,* 22;279(1749): 5048-57. doi: 10.1098/rspb.2012.1108.

Kwon, Y., Kim, S. H., Ronderos, D. S., Lee, Y., Akitake, B., Woodward, O. M., ... and Montell, C. (2010). < i> Drosophila</i> TRPA1 Channel Is Required to Avoid the Naturally Occurring Insect Repellent Citronellal. *Current Biology*, 20(18), 1672-1678.

Launay, P., Fleig, A., Perraud, A. L., Scharenberg, A. M., Penner, R., and Kinet, J. P. (2002). TRPM4 Is a Ca2+-Activated Nonselective Cation Channel Mediating Cell Membrane Depolarization. *Cell,* 109(3), 397-407.

Leypold, B. G., Yu, C. R., Leinders-Zufall, T., Kim, M. M., Zufall, F., and Axel, R. (2002). Altered sexual and social behaviors in trp2 mutant mice. *Proceedings of the National Academy of Sciences*, 99(9), 6376-6381.

Liao, M., Cao, E., Julius, D., and Cheng, Y. (2013). Structure of the TRPV1 ion channel determined by electron cryo-microscopy. *Nature,* 504(7478), 107-112.

Liman, E. R., Corey, D. P., Dulac, C. (1999) TRP2: a candidate transduction channel for mammalian pheromone sensory signaling. *Proc. Natl. Acad. Sci. USA,* 96:5791–5796.

Liman, E. R., and Dulac, C. (2007). TRPC2 and the molecular biology of pheromone detection in mammals. In W. B. Liedtke and S. Heller. TRP ion channel function in sensory transduction and cellular signaling cascades. Boca Raton (Fl): CRC Press. Chapter 3.

Lin, W., Margolskee, R., Donnert, G., Hell, S. W., and Restrepo, D. (2007). Olfactory neurons expressing transient receptor potential channel M5 (TRPM5) are involved in sensing semiochemicals. *Proceedings of the National Academy of Sciences of the United States of America*, 104(7), 2471-2476.

Liu, D., and Liman, E. R. (2003). Intracellular Ca^{2+} and the phospholipid PIP2 regulate the taste transduction ion channel TRPM5. *Proceedings of the National Academy of Sciences,* 100(25), 15160-15165.

Lopez, F., Delgado, R., Lopez, R., Bacigalupo, J., and Restrepo, D. (2014). Transduction for pheromones in the main olfactory epithelium is mediated by the Ca^{2+} -activated channel TRPM5. *The Journal of Neuroscience : The Official Journal of the Society for Neuroscience*, 34(9), 3268-3278.

Lucas, P., Ukhanov, K., Leinders-Zufall, T., and Zufall, F. (2003). A Diacylglycerol-Gated Cation Channel in Vomeronasal Neuron Dendrites Is Impaired in TRPC2 Mutant Mice: Mechanism of Pheromone Transduction. *Neuron,* 40(3), 551-561.

McKemy, D. D., Neuhausser, W. M., and Julius, D. (2002). Identification of a cold receptor reveals a general role for TRP channels in thermosensation. *Nature*, 416(6876), 52-58.

Miller, B. A. (2014). TRPC2. In Mammalian Transient Receptor Potential (TRP) Cation Channels (pp. 53-65). Springer Berlin Heidelberg.

Moiseenkova-Bell, V. Y., Stanciu, L. A., Serysheva, I. I., Tobe, B. J., and Wensel, T. G. (2008). Structure of TRPV1 channel revealed by electron cryomicroscopy. *Proceedings of the National Academy of Sciences,* 105(21), 7451-7455.

Montell, C. (2001). Physiology, phylogeny, and functions of the TRP superfamily of cation channels. Science's STKE : Signal Transduction Knowledge Environment, 2001 (90).

Niemeyer, B. A., Suzuki, E., Scott, K., Jalink, K., and Zuker, C. S. (1996). The drosophila light-activated conductance is composed of the two channels TRP and TRPL. *Cell,* 85(5), 651-659.

Oberwinkler, J., and Philipp, S. E. (2007). Trpm3. In Transient Receptor Potential (TRP) Channels (pp. 253-267). Springer Berlin Heidelberg.

Oshimoto, A., Wakabayashi, Y., Garske, A., Lopez, R., Rolen, S., Flowers, M., Restrepo, D. (2013). Potential role of transient receptor potential channel M5 in sensing putative pheromones in mouse olfactory sensory neurons. *PloS One,* 8(4), 61990.

Peng, G., Shi, X., and Kadowaki, T. (2014). Evolution of TRP channels inferred by their classification in diverse animal species. Molecular phylogenetics and evolution.

Runnels, L. W., Yue, L., and Clapham, D. E. (2001). TRP-PLIK, a bifunctional protein with kinase and ion channel activities. *Science,* 291(5506), 1043-1047.

Saunders, C. J., Christensen, M, Finger, T. E., and Tizzano, M. (2014). Cholinergic Neurotransmission Links Solitary Chemosensory Cells to Nasal Inflammation. *Proc. Natl. Acad. Sci. USA*, 107(7):3210–3215.

Seddon, A. M., Curnow, P., and Booth, P. J. (2004). Membrane proteins, lipids and detergents: not just a soap opera. *Biochimica et Biophysica Acta (BBA)-Biomembranes,* 1666(1), 105-117.

Shpak, G., Zylbertal, A., Yarom, Y., and Wagner, S. (2012). Calcium-activated sustained firing responses distinguish accessory from main olfactory bulb mitral cells. *The Journal of Neuroscience : The Official Journal of the Society for Neuroscience*, 32(18), 6251-6262.

Silver, W. L., P. Roe and C. J. Saunders. (2010) Functional Neuroanatomy of the Upper Airway in Experimental Animals. In Toxicology of the Nose and Upper Airways. Editors J. B. Morris and D. J. Shusterman. pp 45-64.

Störtkuhl, K. F., Hovemann, B. T., Carlson, J. R. (1999). Olfactory adaptation depends on the TRP Ca2+ channel in Drosophila. *Journal of Neuroscience,* 19(12), 4839-4846.

Syam, N., Rougier, J. S., and Abriel, H. (2014). Glycosylation of TRPM4 and TRPM5 channels: molecular determinants and functional aspects. Frontiers in cellular neuroscience, 8.

Thompson, J. A., Salcedo, E., Restrepo, D., and Finger, T. E. (2012). Second-order input to the medial amygdala from olfactory sensory neurons expressing the transduction channel TRPM5. *The Journal of Comparative Neurology*, 520(8), 1819-30.

Vennekens, R., and Nilius, B. (2007). Insights into TRPM4 function, regulation and physiological role. In Transient Receptor Potential (TRP) Channels (pp. 269-285). Springer Berlin Heidelberg.

Voets, T., Droogmans, G., Wissenbach, U., Janssens, A., Flockerzi, V., and Nilius, B. (2004). The principle of temperature-dependent gating in cold- and heat-sensitive TRP channels. *Nature,* 430(7001), 748-754.

Voets, T., Janssens, A., Droogmans, G., and Nilius, B. (2004). Outer pore architecture of a Ca2+-selective TRP channel. *Journal of Biological Chemistry,* 279(15), 15223-15230.

Wehner, F. (2006) Cell volume-regulated cation channels. *Contrib. Nephrol.,* 152:25-53.

Woo, J. S., Lee, K. J., Huang, M., Cho, C. H., Lee, E. H. (2014) Heteromeric TRPC3 with TRPC1 formed via its ankyrin repeats regulates the resting cytosolic Ca2+ levels in skeletal muscle. *Biochem. Biophys. Res. Commun.,* 4;446(2):454-9. doi: 10.1016/j.bbrc.2014.02.127.

Woo, S. K., Kwon, M. S., Ivanov, A., Geng, Z., Gerzanich, V., and Simard, J. M. (2013). Complex N-glycosylation stabilizes surface expression of transient receptor potential melastatin 4b protein. *Journal of Biological Chemistry,* 288(51), 36409-36417.

Yamaguchi, H., Matsushita, M., Nairn, A. C., Kuriyan, J. (2001), Crystal Structure of the Atypical Protein Kinase Domain of a TRP Channel with Phosphotransferase Activity, *Molecular Cell,* 7 (5), 1047-1057.

Yekkirala, A. S. (2013). Two to tango: GPCR oligomers and GPCR-TRP channel interactions in nociception. *Life Sciences*, 92(8), 438-445.

Yildirim, E., Dietrich, A., and Birnbaumer, L. (2003). The mouse C-type transient receptor potential 2 (TRPC2) channel: alternative splicing and calmodulin binding to its N terminus. *Proceedings of the National Academy of Sciences,* 100(5), 2220-2225.

Yoshida, T., Inoue, R., Morii, T., Takahashi, N., Yamamoto, S., Hara, Y., ... and Mori, Y. (2006). Nitric oxide activates TRP channels by cysteine S-nitrosylation. *Nature chemical biology*, 2(11), 596-607.

Young, J. M., Massa, H. F., Hsu, L., and Trask, B. J. (2010). Extreme variability among mammalian V1R gene families. *Genome Research,* 20(1), 10-18.

Yu F. H., and Catterall W. A. (2004) The VGL-chanome: a protein superfamily specialized for electrical signaling and ionic homeostasis. *Sci. STKE,* pp. re15.

Zhang, J. (2003) Evolution by gene duplication: an update. *TRENDS in Ecology and Evolution,* Vol.18 No.6 p 292-298.

Zufall, F. (2005). The TRPC2 ion channel and pheromone sensing in the accessory olfactory system. *Naunyn-Schmiedeberg's Archives of Pharmacology,* 371(4), 245-250.

Zufall, F., Ukhanov, K., Lucas, P., and Leinders-Zufall, T. (2005). Neurobiology of TRPC2: From gene to behavior. *Pflügers Archiv,* 451(1), 61-71.

In: Animal Communication and Cognition ISBN: 978-1-63482-411-8
Editor: Tabitha Wagner

Chapter 4

THE ROLE OF CUTICULAR HYDROCARBON-BASED CHEMICAL COMMUNICATION IN PREZYGOTIC REPRODUCTIVE ISOLATION

***J. Buellesbach*[1,2,3,*] *and T. Schmitt*[1,2,4]**

[1]Faculty of Biology, Department of Evolutionary Biology and Animal Ecology, University of Freiburg, Freiburg, Germany
[2]Spemann Graduate School of Biology and Medicine (SGBM), University of Freiburg, Freiburg, Germany
[3]Jena School for Microbial Communication (JSMC), Friedrich Schiller University Jena, Jena, Germany
[4]Faculty of Biology, Department of Animal Ecology and Tropical Biology, University of Würzburg, Würzburg, Germany

One of the most important questions in evolutionary biology is how new species originate. In the last decades, research on speciation has focused almost entirely on the development and maintenance of barriers to interspecific reproduction (i.e. hybridization), postulated to be the major driving force of speciation. Prezygotic reproductive isolation includes mechanisms contributing to ecological, behavioral and mechanical isolation between species, leading to different degrees of gene flow reduction or complete interruption. Therefore, any cues or signals capable of accurately

* Corresponding author: J. Buellesbach. Jena School for Microbial Communication (JSMC), Friedrich Schiller University Jena, Jena, Germany. E-mail: jan.buellesbach@web.de.

conveying information about species, sex and reproductive status have the potential to contribute to prezygotic reproductive isolation between diverging species. Albeit studies on animal signals have been mostly devoted to visual and auditory communication, empirical evidence has accumulated that olfactory communication (i.e. chemical signaling) also plays a major role in animal communication systems. In particular, insects have exploited this type of signaling as their primary communication mode. In the following chapter, we present an overview over the contribution of chemical communication to prezygotic reproductive isolation in exemplary insect species with a particular emphasis on the variation of cuticular hydrocarbons between species and its impact on mate preference and assortative mating.

A recent study strongly suggests that the complex olfactory system of insects evolved long after their shift from marine to terrestrial territory as an adaptation to their ability to fly (Missbach et al. 2014). Hence, over the long period of adaptation to a terrestrial and partially airborne life style, chemical signaling gradually developed as the primary communication mode in insects (Greenfield 2002). Cues and signals involved in chemical communication capable of modifying the receivers' behavior are generally called semiochemicals (Regnier 1971). These can be further divided according to their function in either inter- or intraspecific chemical communication (Regnier 1971; Sonenshine 1985). In interspecific communication, two main types of semiochemicals can be differentiated, ones that are adaptively favorable to the sender (i.e. allomones) and ones that are adaptively favorable to the receiver (i.e. kairomones) (Brown et al. 1970; Regnier 1971). In intraspecific communication, chemical signals exchanged between conspecific individuals are generally defined as pheromones (Karlson and Lüscher 1959). The most thoroughly investigated pheromones are sex attractants (Karlson and Butenandt 1959), fundamental elements in sexual communication typically functioning to attract conspecific mating partners (e.g. Baker 1989; Wyatt 2003) and to signal fertility, reproductive status and mate quality (e.g. Hardie and Minks 1999; Johansson and Jones 2007).

Females (e.g. Antony and Jallon 1982; Löfstedt 1993; Wyatt 2003) as well as males (e.g. Grillet et al. 2006; Ruther et al. 2007; Eltz et al. 2008) can be the emitters of sex attractants, imposing the role of signal perception on the respective other sex. Both parts of the communication system are required to be tightly coordinated to ensure accurate and successful mutual attraction of mating partners (Bargmann 2006). Pheromone blends and the capabilities to perceive and discriminate between them often differ considerably between species, resulting in distinct, species-specific "communication channels" ensuring exclusive intraspecific sexual communication (Paterson 1980; Baker 1989). Hence, these communication channels have the potential to reproductively isolate species occurring in sympatry which show distinctive enough signaling and perception capabilities (Löfstedt 1993; Bargmann 2006). Intriguingly, comparative studies of species-specific sex pheromone blends between closely related sympatric species reveal rather striking differences, strongly hinting at increased signal divergence in sympatry (e.g. Löfstedt et al. 1991; McElfresh and Millar 1999; Gries et al. 2001). Concordantly, additional evidence stemming from behavioral experiments shows that sympatry apparently also enhances discrimination behavior against heterospecific mating partners, where clear preference behavior based on conspecific sexual cues could be demonstrated (e.g. Coyne and Orr 1989, 1997). Attraction and preference of mating partners has been shown to be tightly associated with their species-specific pheromone blends for instance in moths of the genus *Archips* (Carde et al. 1977) and in parasitic wasps of the genus *Glyptapanteles* (Danci et al. 2006). These examples emphasize the considerable potential of species- and sex-specific chemical cues and signals to promote and preserve prezygotic reproductive isolation between diverging species (Johansson and Jones 2007; Smadja and Butlin 2009). A notable challenge in these studies is the identification and subsequent isolation of the particular semiochemicals functioning as the main cues or signals which elicit the discriminatory behavior ultimately contributing to prezygotic reproductive isolation.

Cuticular hydrocarbons (CHC), the dominant fraction of the waxy lipid layer located on the insect epicuticle (Howard and Blomquist 1982; Lockey 1985), form the basis for a wide array of semiochemicals involved in a large variety of signaling systems (Blomquist and Bagneres 2010). For instance, it has been demonstrated that CHC can function as dominance and fertility cues (e.g. Monnin et al. 1998; Liebig et al. 2000), as territorial and thermoregulatory signals (e.g. Howard and Blomquist 1982; 2005) or in chemical mimicry (e.g. Dettner and Liepert 1994; Lenoir et al. 2001; Strohm et al. 2008). Moreover, in eusocial Hymenoptera, i.e. ants as well as certain bee and wasp species, CHC are fundamental as the major nestmate and caste recognition cues (e.g. Van der Meer and Werren 1998; Breed 2003; Gamboa 2004; Dani et al. 2005). But most prominently, CHC have been shown to play pivotal roles in sexual communication as sex attractants, constituting the main cues to attract conspecific mates (Antony and Jallon 1982; Cobb and Jallon 1990) to enable discrimination of con- from heterospecific mating partners (e.g. Singer 1998; Howard et al. 2003) and to signal receptiveness and fertility (Simmons et al. 2003).

In order to function in intraspecific sexual signaling, sexual dimorphism is an important prerequisite of CHC profiles, since potential mating partners need to be able to recognize and distinguish the respective other sex (Howard and Blomquist 2005). It has been shown that intraspecific sex-specificity of CHC profiles appears to be widely distributed, but not universal characteristic across different insect taxa (Thomas and Simmons 2008). In fact, the presence or absence of sexual dimorphism in CHC profiles can constitute a distinct character trait differentiating closely related species, as demonstrated e.g. in the eight species of the *Drosophila melanogaster* subgroup (Jallon and David 1987, see also below).

Furthermore, to function as species-specific recognition cues, CHC profiles need to be distinctive enough to separate different species (Singer 1998). Several studies found CHC profiles between closely related species to be sufficiently distinct for their unambiguous characterization and separation (e.g. Carlson and Service 1979; Howard et al. 1988; Page et al. 1997). Species-specificity of CHC profiles has been further demonstrated in comparisons between closely related species of the Coleopteran families Curculionidae (Page et al. 1997) and Scolytidae (Page et al. 1990a, b), the Isopteran families Rhinotermitidae (Bagneres et al. 1991; Takahashi and Gassa 1995), Termitidae (Howard et al. 1988) and Termopsidae (Haverty et al. 1990), in the Dipteran families Culicidae (Carlson and Service 1979, 1980), Muscidae (Nelson and Carlson 1986), Simulidae (Carlson and Walsh 1981) and Drosophilidae (Ferveur 2005) and finally in the Hymenopteran families Formicidae (Martin and Drijfhout 2009), Megachilidae (Buckner et al. 2009), Apidae (Hadley et al. 1981), Vespidae (Butts et al. 1991), Bethylidae (Howard 1992) and Pteromalidae (Bernier et al. 1998). Species and sex specificity could function as prezygotic hybridization barrier when the receiving sex is able to discriminate between a con- and a heterospecific mating partner. However, this has been rarely proven referring to CHC profiles as sexual cues.

Only if both prerequisites, sexual dimorphism and species-specificity, are met then CHC profiles bear the most potential to function as mate discrimination cues in sexual signaling and concordantly, to contribute to prezygotic reproductive isolation. Curiously, relatively few studies have fully explored the potential of diverging CHC profiles to prezygotic reproductive isolation between closely related species. This is most likely due to the paucity of suitable, closely related groups of model organisms, restricted feasibility of cross-specific behavioral assays and complexity of comparative chemical analyses. Whereas these aspects render an empirical demonstration of the contribution of CHC profile divergence to reproductive isolation difficult, several studies provided evidence for the involvement of CHC-based sex- and species-specific signaling in prezygotic reproductive isolation. For instance, mate choice and preference behavior between closely related species in the Coleopteran families Cerambycidae (Tanigaki et al. 2007), Chrysomelidae (Peterson et al. 2007) and Lampyridae (South et al. 2008), appears to be mediated through CHC-based sexual signaling, as choice experiments have clearly demonstrated.

Contrasting the above-mentioned single case studies, the fruit fly genus *Drosophila* has been particularly well studied concerning species-specific CHC variation as well as mechanisms of CHC-based mate recognition and species discrimination (e.g. Buckley et al. 1997; Ferveur 1997; Ferveur 2005). For different *Drosophila* species, CHC profiles appear to be either sexually mono- or dimorphic (Jallon and David 1987). In dimorphic species like *D. melanogaster*, distinct qualitative and quantitative differences between males and females clearly separate their CHC profiles (Jallon 1984; Jallon and David 1987). Males predominantly produce alkenes, whereas alkadienes are specifically found in females (Jallon 1984). Particular alkadiene compounds from female CHC profiles can serve as sex attractants for conspecific males (Antony et al. 1985; Ferveur and Sureau 1996), while male alkenes have been shown to stimulate conspecific females (Grillet et al. 2006). Conversely, in sexually monomorphic species like *D. simulans*, CHC profiles do not show any detectable differences between the sexes (Jallon 1984; Jallon and David 1987). It has been shown that this CHC difference contributes to the prevention of cross-attraction between sexually mono- and dimorphic species (Coyne et al. 1994; Coyne and Charlesworth 1997; Carracedo et al. 2003). Moreover, between certain *Drosophila* species occurring both allopatrically and sympatrically, CHC differences as well as CHC-mediated mate discrimination appear to be more pronounced in the latter case, suggesting stronger selection for both divergent CHC profiles and the perception capabilities to discriminate between them if they share the same habitat (Ishii et al. 2001; Mas and Jallon 2005; Higgie and Blows 2007).

In Hymenoptera, on the other hand, extensive research has been devoted to the phenomenon of CHC-based nestmate and caste recognition (see above), but this is limited to eusocial species where reproduction is restricted to only one or a few individuals per colony, greatly altering the modalities of sexual signaling. Concerning solitary Hymenopteran species, however, recent advances have been made in understanding CHC-based sexual communication in the jewel wasp species complex *Nasonia*, an emerging model system for investigating ecological (e.g. Werren 1980, 1983), behavioral (e.g. van den

Assem and Werren 1994; Leonard and Boake 2006) and genetic (e.g. Breeuwer and Werren 1995; Pultz and Leaf 2003) aspects of reproductive isolation. Carlson et al. (1999) were the first to investigate CHC profiles of the most prominent and cosmopolitan member of the *Nasonia* genus, *N. vitripennis*, and found clear sex-specific differences between adult males and females. Furthermore, Steiner et al. (2006) found experimental evidence that female CHC profiles function as sex pheromones in *N. vitripennis*, initiating courtship behavior in conspecific males. Species-specificity in *Nasonia* CHC profiles was further demonstrated to be pronounced enough to discriminate the two most recently diverged sister species, *N. oneida* and *N. giraulti* (Raychoudhury et al. 2010). Extending the CHC profile comparison to all four members of the *Nasonia* species complex, it has been demonstrated that they are sufficiently distinct to unambiguously separate the *Nasonia* genus according to sex and species (Buellesbach et al. 2013). Moreover, the species-specific variation in female CHC profiles contributes to male preference and discrimination of con- over heterospecific mating partners, albeit differentially pronounced in the different species (Buellesbach et al. 2013; 2014). These findings strongly suggest the involvement of female CHC as species-specific female-produced sexual cues in prezygotic reproductive isolation in the *Nasonia* species complex. Extrapolating these findings to more distant relatives of *Nasonia* in their family of chalcidoid parasitoid wasps (Pteromalidae) hint at an early evolution of sex-specific CHC differences and their potential utilization as female sexual cues in this taxa (Buellesbach et al., unpublished data).

In conclusion, sex-specific chemical cues distinctive enough to separate different closely related species bear huge potential to contribute to prezygotic reproductive isolation if they function in well-separated sexual communication channels. Further efforts should be directed towards identifying and studying more suitable model organisms to investigate this phenomenon, as well as extending the studies to other classes of semiochemicals with the potential to be involved in and contribute to prezygotic reproductive isolation.

References

Antony, C., T. L. Davis, D. A. Carlson, J. M. Pechine, and J. M. Jallon. 1985. Compared behavioral responses of male *Drosophila melanogaster* (Canton-S) to natural and synthetic aphrodisiacs. *J. Chem. Ecol.* 11:1617-1629.

Antony, C. and J. M. Jallon. 1982. The chemical basis for sex recognition in *Drosophila melanogaster*. *J. Insect Physiol.* 28:873-880.

Bagneres, A. G., A. Killian, J. L. Clement, and C. Lange. 1991. Interspecific recognition among termites of the genus *Reticulitermes*: Evidence for a role for the cuticular hydrocarbons. *J. Chem. Ecol.* 17:2397-2420.

Baker, T. C. 1989. Sex-pheromone communication in the Lepidoptera: New research progress. *Experientia* 45:248-262.

Bargmann, C. I. 2006. Comparative chemosensation from receptors to ecology. *Nature* 444:295-301.

Bernier, U. R., D. A. Carlson and C. J. Geden. 1998. Gas chromatography/ mass spectrometry analysis of the cuticular hydrocarbons from parasitic wasps of the genus *Muscidifurax*. *J. Am. Soc. Mass Spectrom.* 9:320-332.

Blomquist, G. J. and A. G. Bagneres. 2010. *Insect Hydrocarbons: Biology, Biochemistry, and Chemical Ecology,* Cambridge University Press, New York, US.

Breed, M. D. 2003. Nestmate recognition assays as a tool for population and ecological studies in eusocial insects: A review. *J. Kans. Entomol. Soc.* 76:539-550.

Breeuwer, J. A. J. and J. H. Werren. 1995. Hybrid breakdown between two haplodiploid species: The role of nuclear and cytoplasmic genes. *Evolution* 49:705-717.

Brown, W. L., T. Eisner and R. H. Whittake. 1970. Allomones and kairomones: Transspecific chemical messengers. *BioScience* 20:21-22.

Buckley, S. H., T. Tregenza and R. K. Butlin. 1997. Speciation and signal trait genetics. *Trends Ecol. Evol.* 12:299-301.

Buckner, J. S., T. L. Pitts-Singer, C. Guedot, M. M. Hagen, C. L. Fatland, and W. P. Kemp. 2009. Cuticular lipids of female solitary bees, *Osmia lignaria* Say and *Megachile rotundata* (F.) (Hymenoptera: Megachilidae). *Comp. Biochem. Phys. B* 153:200-205.

Buellesbach, J., J. Gadau, L. W. Beukeboom, F. Echinger, R. Raychoudhury, J. H. Werren, and T. Schmitt. 2013. Cuticular hydrocarbon divergence in the jewel wasp *Nasonia*: evolutionary shifts in chemical communication channels? *J. Evol. Biol.* 26:2467-2478.

Buellesbach, J., C. Greim and T. Schmitt. 2014. Asymmetric interspecific mating behavior reflects incomplete prezygotic isolation in the jewel wasp genus *Nasonia*. *Ethology* 120:834-843.

Butts, D. P., K. E. Espelie and H. R. Hermann. 1991. Cuticular hydrocarbons of 4 species of social wasps in the subfamily Vespinae - *Vespa crabro* L,

Dolichovespula maculata (L), *Vespula squamosa* (Drury), and *Vespula maculifrons* (Buysson). *Comp. Biochem. Phys. B* 99:87-91.

Carde, R. T., A. M. Carde, A. S. Hill, and W. L. Roelofs. 1977. Sex pheromone specificity as a reproductive isolating mechanism among sibling species *Archips argyrospilus* and *A. mortuanus* and other sympatric tortricine moths (Lepidoptera: Tortricidae). *J. Chem. Ecol.* 3: 71-84.

Carlson, D. A., C. J. Geden and U. R. Bernier. 1999. Identification of pupal exuviae of *Nasonia vitripennis* and *Muscidifurax raptorellus* parasitoids using cuticular hydrocarbons. *Biol. Control* 15:97-105.

Carlson, D. A. and M. W. Service. 1979. Differentiation between species of the Anopheles gambiae giles complex (Diptera, Culicidae) by analysis of cuticular hydrocarbons. *Ann. Trop. Med. Parasitol.* 73:589-592.

Carlson, D. A. and M. W. Service. 1980. Identification of mosquitos of *Anopheles gambiae* species complex A and complex B by analysis of cuticular components. *Science* 207:1089-1091.

Carlson, D. A. and J. F. Walsh. 1981. Identification of 2 west African Black flies (Diptera, Simuliidae) of the Simulium-Damnosum species complex by analysis of cuticular paraffins. *Acta Trop.* 38:235-239.

Carracedo, M. C., A. Asenjo and P. Casares. 2003. Genetics of *Drosophila simulans* male mating discrimination in crosses with *D. melanogaster. Heredity* 91:202-207.

Cobb, M. and J. M. Jallon. 1990. Pheromones, mate recognition and courtship stimulation in the *Drosophila melanogaster* species subgroup. *Anim. Behav.* 39:1058-1067.

Coyne, J. A. and B. Charlesworth. 1997. Genetics of a pheromonal difference affecting sexual isolation between *Drosophila mauritiana* and *D. sechellia*. *Genetics* 145:1015-1030.

Coyne, J. A., A. P. Crittenden and K. Mah. 1994. Genetics of a pheromonal difference contributing to reproductive isolation in *Drosophila*. *Science* 265:1461-1464.

Coyne, J. A. and H. A. Orr. 1989. Patterns of speciation in *Drosophila*. *Evolution* 43:362-381.

Coyne, J. A. and H. A. Orr. 1997. "Patterns of speciation in *Drosophila*" revisited. *Evolution* 51:295-303.

Danci, A., P. W. Schaefer A. Schopf, and G. Gries. 2006. Species-specific close-range sexual communication systems prevent cross-attraction in three species of *Glyptapanteles* parasitic wasps (Hymenoptera: Braconidae). *Biol. Control* 39:225-231.

Dani, F. R., G. R. Jones, S. Corsi, R. Beard, D. Pradella, and S. Turillazzi. 2005. Nestmate recognition cues in the honey bee: Differential importance of cuticular alkanes and alkenes. *Chem. Senses* 30:477-489.

Dettner, K. and C. Liepert. 1994. Chemical mimicry and camouflage. *Annu. Rev. Entomol.* 39:129-154.

Eltz, T., Y. Zimmermann, C. Pfeiffer, J. R. Pech, R. Twele, W. Francke, J. J. G. Quezada-Euan, and K. Lunau. 2008. An olfactory shift is associated with male perfume differentiation and species divergence in orchid bees. *Curr. Biol.* 18:1844-1848.

Ferveur, J.-F. 2005. Cuticular hydrocarbons: Their evolution and roles in *Drosophila* pheromonal communication, *Behav. Genet.* 35:279-295.

Ferveur, J. F. 1997. The pheromonal role of cuticular hydrocarbons in *Drosophila melanogaster*. *BioEssays* 19:353-358.

Ferveur, J. F. and G. Sureau. 1996. Simultaneous influence on male courtship of stimulatory and inhibitory pheromones produced by live sex-mosaic *Drosophila melanogaster*. *Proc. R. Soc. Lond. B Biol. Sci.* 263:967-973.

Gamboa, G. J. 2004. Kin recognition in eusocial wasps. *Ann. Zool. Fenn.* 41: 789-808.

Greenfield, M. D. 2002. *Signalers and Receivers: Mechanisms and Evolution of Arthropod Communication*. Oxford University Press, New York, US.

Gries, G., P. W. Schaefer, R. Gries, J. Liska, and T. Gotoh. 2001. Reproductive character displacement in *Lymantria monacha* from northern Japan? *J. Chem. Ecol.* 27:1163-1176.

Grillet, M., L. Dartevelle and J.-F. Ferveur. 2006. A *Drosophila* male pheromone affects female sexual receptivity. *Proc. R. Soc. Lond. B Biol. Sci.* 273:315-323.

Hadley, N. F., G. J. Blomquist and U. N. Lanham. 1981. Cuticular hydrocarbons of 4 species of colorado hymenoptera. *Insect Biochem.* 11: 173-177.

Hardie, J. and A. K. Minks. 1999. *Pheromones of non-Lepidopteran insects associated with agricultural plants*. CABI Press, Wallingford, UK.

Haverty, M. I., L. J. Nelson and M. Page. 1990. Cuticular hydrocarbons of 4 populations of *Coptotermes formosanus* Shiraki (Isoptera, Rhinotermitidae) in the United States: Similarities and origins of introductions. *J. Chem. Ecol.* 16:1635-1647.

Higgie, M. and M. W. Blows. 2007. Are traits that experience reinforcement also under sexual selection? *Am. Nat.* 170:409-420.

Howard, R. W. 1992. Comparative-analysis of cuticular hydrocarbons from the ectoparasitoids Cephalonomia waterstoni and Laelius utilis

(Hymenoptera, Bethylidae) and their respective hosts, Cryptolestes ferrugineus (Coleoptera, Cucujidae) and Trogoderma variabile (Coleoptera, Dermestidae). *Ann. Entomol. Soc. Am.* 85:317-325.

Howard, R. W. and C. Blomquist. 1982. Chemical ecology and biochemistry of insect hydrocarbons. *Annu. Rev. Entomol.* 27:149-172.

Howard, R. W. and G. J. Blomquist. 2005. Ecological, behavioral, and biochemical aspects of insect hydrocarbons. *Annu. Rev. Entomol.* 50:371-393.

Howard, R. W., L. L. Jackson, H. Banse, and M. W. Blows. 2003. Cuticular hydrocarbons of *Drosophila birchii* and *D. serrata*: Identification and role in mate choice in *D. serrata*. *J. Chem. Ecol.* 29:961-976.

Howard, R. W., B. L. Thorne, S. C. Levings, and C. A. McDaniel. 1988. Cuticular hydrocarbons as chemotaxonomic characters for *Nasutitermes corniger* (Motschulsky) and *Nasutitermes ephratae* (Holmgren) (Isoptera, Termitidae). *Ann. Entomol. Soc. Am.* 81:395-399.

Ishii, K., Y. Hirai, C. Katagiri, and M. T. Kimura. 2001. Sexual isolation and cuticular hydrocarbons in *Drosophila elegans*. *Heredity* 87:392-399.

Jallon, J. M. 1984. A few chemical words exchanged by *Drosophila* during courtship and mating. *Behav. Genet.* 14:441-478.

Jallon, J. M. and J. R. David. 1987. Variations in cuticular hydrocarbons among the 8 species of the *Drosophila melanogaster* subgroup. *Evolution* 41:294-302.

Johansson, B. G. and T. M. Jones. 2007. The role of chemical communication in mate choice. *Biol. Rev.* 82:265-289.

Karlson, P. and A. Butenandt. 1959. Pheromones (ectohormones) in insects. *Annu. Rev. Entomol.* 4:39-58.

Karlson, P. and M. Lüscher. 1959. Pheromones: A new term for a class of biologically active substances. *Nature* 183:55-56.

Lenoir, A., P. D'Ettorre, C. Errard, and A. Hefetz. 2001. Chemical ecology and social parasitism in ants. *Annu. Rev. Entomol.* 46:573-599.

Leonard, J. E. and C. R. B. Boake. 2006. Site-dependent aggression and mating behaviour in three species of *Nasonia* (Hymenoptera: Pteromalidae). *Anim. Behav.* 71:641-647.

Liebig, J., C. Peeters, N. J. Oldham, C. Markstadter, and B. Holldobler. 2000. Are variations in cuticular hydrocarbons of queens and workers a reliable signal of fertility in the ant *Harpegnathos saltato*r? *Proc. Natl. Acad. Sci. US* 97:4124-4131.

Lockey, K. H. 1985. Insect cuticular lipids. *Comp. Biochem. Phys. B* 81:263-273.

Löfstedt, C. 1993. Moth pheromone genetics and evolution. *Philos. Trans. R. Soc. Lond. B Biol. Sci.* 340:167-177.

Löfstedt, C., W. M. Herrebout and S. B. J. Menken. 1991. Sex pheromones and their potential role in the evolution of reproductive isolation in small ermine moths (Yponomeutidae). *Chemoecology* 2:20-28.

Martin, S. and F. Drijfhout. 2009. A review of ant cuticular hydrocarbons. *J. Chem. Ecol.* 35:1151-1161.

Mas, F. and J. M. Jallon. 2005. Sexual isolation and cuticular hydrocarbon differences between *Drosophila santomea* and *Drosophila yakuba*. *J. Chem. Ecol.* 31:2747-2752.

McElfresh, J. S. and J. G. Millar. 1999. Geographic variation in sex pheromone blend of *Hemileuca electra* from Southern California. *J. Chem. Ecol.* 25:2505-2525.

Missbach, C., H. K. Dweck, H. Vogel, A. Vilcinskas, M. C. Stensmyr, B. S. Hansson, and E. Grosse-Wilde. 2014. *Evolution of insect olfactory receptors.*

Monnin, T., C. Malosse and C. Peeters. 1998. Solid-phase microextraction and cuticular hydrocarbon differences related to reproductive activity in queenless ant *Dinoponera quadriceps*. *J. Chem. Ecol.* 24:1423-1423.

Nelson, D. R. and D. A. Carlson. 1986. Cuticular hydrocarbons of the tsetse flies *Glossina morsitans morsitans*, *Glossina austeni* and *Glossina pallidipes*. *Insect Biochem.* 16:403-416.

Page, M., L. J. Nelson, G. J. Blomquist, and S. J. Seybold. 1997. Cuticular hydrocarbons as chemotaxonomic characters of pine engraver beetles (*Ips* spp.) in the *grandicollis* subgeneric group. *J. Chem. Ecol.* 23:1053-1099.

Page, M., L. J. Nelson, M. I. Haverty, and G. J. Blomquist. 1990a. Cuticular hydrocarbons as chemotaxonomic characters for bark beetles: *Dendroctonus ponderosae*, *Dendroctonus jeffreyi*, *Dendroctonus brevicomis*, and *Dendroctonus frontalis* (Coleoptera, Scolytidae). *Ann. Entomol. Soc. Am.* 83:892-901.

Page, M., L. J. Nelson, M. I. Haverty, and G. J. Blomquist. 1990b. Cuticular hydrocarbons of 8 species of North American cone beetles, *Conophthorus* Hopkins. *J. Chem. Ecol.* 16:1173-1198.

Paterson, H. E. 1980. A comment on mate recognition systems. *Evolution* 34: 330-331.

Peterson, M. A., S. Dobler, E. L. Larson, D. Juarez, T. Schlarbaum, K. J. Monsen, and W. Francke. 2007. Profiles of cuticular hydrocarbons mediate male mate choice and sexual isolation between hybridising *Chrysochus* (Coleoptera : Chrysomelidae). *Chemoecology* 17:87-96.

Pultz, M. A. and D. S. Leaf. 2003. The jewel wasp *Nasonia*: Querying the genome with haplo-diploid genetics. *Genesis* 35:185-191.

Raychoudhury, R., C. A. Desjardins, J. Buellesbach, D. W. Loehlin, B. K. Grillenberger, L. W. Beukeboom, T. Schmitt, and J. H. Werren. 2010. Behavioral and genetic characteristics of a new species of *Nasonia*. *Heredity* 104:278-288.

Regnier, F. E. 1971. Semiochemicals - Structure and function. *Biology of Reproduction* 4:309-326.

Ruther, J., L. M. Stahl, S. Steiner, L. A. Garbe, and T. Tolasch. 2007. A male sex pheromone in a parasitic wasp and control of the behavioral response by the female's mating status. *J. Exp. Biol.* 210:2163-2169.

Simmons, L. W., J. Alcock and A. Reeder. 2003. The role of cuticular hydrocarbons in male attraction and repulsion by female Dawson's burrowing bee, *Amegilla dawsoni*. *Anim. Behav.* 66:677-685.

Singer, T. L. 1998. Roles of hydrocarbons in the recognition systems of insects. *Am. Zool.* 38:394-405.

Smadja, C. and R. K. Butlin. 2009. On the scent of speciation: The chemosensory system and its role in premating isolation. *Heredity* 102:77-97.

Sonenshine, D. E. 1985. Pheromones and other semiochemicals of the Acari. *Annual review of Entomology* 30:1-28.

South, A., K. Levan, L. Leombruni, C. M. Orians, and S. M. Lewis. 2008. Examining the role of cuticular hydrocarbons in firefly species recognition. *Ethology* 114:916-924.

Steiner, S., N. Hermann and J. Ruther. 2006. Characterization of a female-produced courtship pheromone in the parasitoid *Nasonia vitripennis*. *J. Chem. Ecol.* 32:1687-1702.

Strohm, E., J. Kroiss, G. Herzner, C. Laurien-Kehnen, W. Boland, P. Schreier, and T. Schmitt. 2008. A cuckoo in wolves' clothing? Chemical mimicry in a specialized cuckoo wasp of the European beewolf (Hymenoptera, Chrysididae and Crabronidae). *Front. Zool.* 5.

Takahashi, S. and A. Gassa. 1995. Roles of cuticular hydrocarbons in intra- and interspecific recognition behavior of two Rhinotermitidae species. *J. Chem. Ecol.* 21:1837-1845.

Tanigaki, T., R. Yamaoka and T. Sota. 2007. The role of cuticular hydrocarbons in mating and conspecific recognition in the closely related longicorn beetles *Pidonia grallatrix* and *P. takechii*. *Zool. Sci.* 24:39-45.

Thomas, M. L. and L. W. Simmons. 2008. Sexual dimorphism in cuticular hydrocarbons of the Australian field cricket *Teleogryllus oceanicus* (Orthoptera: Gryllidae) *J. Insect Physiol.* 54:1081-1089.

Van den Assem, J. and J. H. Werren. 1994. A comparison of the courtship and mating behavior of three species of *Nasonia* (Hymenoptera: Pteromalidae). *J. Insect Behav.* 7:53-66.

Van der Meer, R. K. and J. H. Werren. 1998. Nestmate recognition in ants. pp. 79-103 In: R. K. Van der Meer, M. Breed, M. Winston, and K. E. Espelie, eds. *Pheromone Communication in Social Insects.* Westview Press, Boulder, Colorado.

Werren, J. H. 1980. Sex ratio adaptions to local mate competition in a parasitic wasp. *Science* 208:1157-1159.

Werren, J. H. 1983. Sex ratio evolution under local mate competition in a parasitic wasp. *Evolution* 37:116-124.

Wyatt, T. D. 2003. *Sex Pheromones: Finding and choosing mates. Pheromones and Animal Behavior. Communication by Smell and Taste.* Cambridge University Press, Cambridge.

INDEX

A

access, 62, 63
action potential, 99
adaptation, 103, 105, 108
adjustment, 51
adults, 62
age, 79, 88
aggression, 4, 44, 50, 91, 97, 116
aggressive behavior, 97
agonist, 94, 95
algorithm, 10, 14, 16, 25
alkenes, 111, 115
American Sign Language, viii, 57, 62
amino, 92
amino acid(s), 92
Amphibolurus muricatus, vii, 2, 49, 51, 52, 53, 54, 55
amplitude, 43, 44
amygdala, 88, 97, 105
anatomy, 88
ancestors, 99
animal behavior, 48, 54, 84, 85
animal communication, vii, ix, 1, 40, 46, 47, 54, 108
animal interactions, vii, 1
animal signals, viii, 48, 52, 108
animations, vii, 2, 6, 7, 16, 17, 19, 33, 36, 39, 40, 41, 49, 50
anthropology, 82
appeasement, 4, 44
artificial stimuli, vii, 1, 21
ASL, 61, 62, 64, 67, 70
assessment, 6, 39, 40, 44, 45, 48, 51, 53
atmosphere, 54
attribution, viii, 57, 65, 66, 67, 71, 73, 75, 76, 77, 78
avian, 25, 50, 54

B

background noise, 52
barriers, viii, 80, 107
base, 68, 73
bedding, 102
beetles, 117, 118
behavioral experiments, vii, 1, 109
behaviors, 40, 54, 58, 60, 63, 64, 65, 70, 75, 80, 88, 98
benefits, 44, 55
bias, 45, 47, 52
biochemistry, 116
biological motion, 40
birds, 46, 48, 50, 53
blends, 99, 108
blood, 101
body size, 52
bone(s), 10, 11, 12, 14, 15
brain, 83, 88, 99, 102
brain size, 83

breakdown, 113
buttons, 68, 73

C

Ca^{2+}, 95, 103, 105
calcium, 89, 90, 95
caregivers, 62, 63, 72, 77
cascades, 103
case studies, 111
categorization, 70
category a, 67
cation, 89, 104, 105
causal relationship, 59
cellular signaling pathway, 90
changing environment, 103
channel blocker, 98
chemical(s), vii, viii, ix, 87, 88, 89, 96, 99, 100, 106, 108, 109, 110, 112, 113, 116
childhood, 81, 84
children, vii, 57, 58, 59, 60, 61, 62, 63, 67, 78, 79, 80, 81, 83, 84, 85
chimpanzee(s), vii, viii, 57, 59, 60, 61, 62, 63, 64, 65, 67, 68, 70, 71, 72, 75, 77, 78, 79, 80, 81, 82, 83, 84, 85
chromatography, 113
clarity, 27
classes, 112
classification, 70, 85, 104
cleaning, 77
clothing, 118
coding, 14, 96
coffee, 75
cognition, 81, 85
cognitive process, 3
coherence, 19, 20, 21, 26, 27, 28, 30, 31, 49
color, 2, 3, 11, 13, 20, 41, 93
communication, vii, viii, 1, 2, 3, 6, 7, 22, 25, 26, 35, 36, 37, 38, 40, 46, 47, 51, 52, 53, 54, 79, 80, 83, 88, 89, 100, 108, 109, 111, 112, 113, 114, 115, 116
communication systems, ix, 47, 108, 114
community, 61, 72, 79, 80
competition, 44, 119
complex interactions, 37
complexity, 33, 41, 44, 46, 51, 59, 110
compounds, 94, 111
computer, vii, 1, 2, 7, 16, 19, 32, 36, 39, 40, 41, 50, 54
computer-animated stimuli, vii, 1, 54
computer-generated animation, vii, 1, 2, 7, 32, 39, 40, 41, 54
conductance, 92, 101, 104
conduction, 101
configuration, 65, 67
conflict, 44
consensus, 51
conservation, 93, 94
containers, 63
conversations, 62, 81
cooking, 58
coordination, 94
copulation, 75
courtship, 3, 47, 52, 112, 114, 115, 116, 118, 119
crystal structure, 102
cues, viii, 5, 16, 20, 21, 27, 36, 44, 49, 52, 53, 55, 107, 109, 110, 112, 115
culture, 81, 84
cuticular hydrocarbon, vii, ix, 108, 113, 114, 115, 116, 117, 118, 119
cysteine, 94, 106

D

database, 63, 64
deduction, 78
deficit, 37
depolarization, viii, 88, 95, 96, 99
depth, 5, 31
detectable, 111
detection, 16, 21, 26, 28, 31, 32, 35, 41, 46, 47, 51, 54, 89, 91, 94, 97, 100, 103
detection system, 100
detergents, 104
dimorphism, 110, 119
dinosaurs, 48
diploid, 118
discrimination, 3, 6, 7, 19, 20, 50, 53, 109, 110, 111, 112, 114

diseases, 100
displacement, 115
distribution, 10, 12
divergence, 46, 50, 52, 90, 109, 110, 113, 115
diversification, 90, 91
diversity, 5, 53, 94, 97
DNA, 90
dominance, 109
Drosophila, viii, 87, 89, 103, 105, 110, 111, 112, 113, 114, 115, 116, 117

E

ecology, 47, 49, 113, 116
ecosystem, 26
electron, 92, 101, 103, 104
electron microscopy, 101
emitters, 108
emotional responses, 88
endocrine, 89
endocrine system, 89
endurance, 46
energy, 50
engineering, 10
environment(s), 3, 14, 16, 19, 28, 48, 50, 51, 52, 60, 62, 63, 72, 75
environmental influences, 48
epithelia, 97
epithelium, 91, 103
evidence, viii, 16, 38, 45, 49, 59, 78, 80, 88, 108, 109, 111, 112
evolution, vii, viii, 2, 3, 6, 19, 21, 22, 26, 41, 43, 45, 46, 47, 48, 50, 51, 52, 53, 54, 81, 84, 88, 89, 99, 104, 112, 115, 117, 119
exploitation, 3, 26
exposure, 41, 79

F

false belief, 84
families, 90, 94, 96, 106, 110, 111
family members, viii, 88, 89, 92, 93, 99
fantasy, 58, 67, 83, 84
fertility, 46, 108, 109, 116
fibers, 100
field tests, 53
fish, 3, 45, 50, 54, 77, 91
flexibility, 7
flight, 53, 88
food, 59
force, viii, 107
formation, 94
fragments, 77
fungi, 90
fusion, 19, 49, 50

G

gene flow, viii, 107
genes, 89, 90, 91, 96, 99, 101, 102, 113
genetic components, 102
genetics, 113, 117, 118
genome, 90, 91, 102, 118
genus, 45, 109, 111, 112, 113
geography, 42
glycosylation, 94, 105
grass, 48
grouping, 49
guidelines, 45

H

habitat(s), 6, 41, 42, 50, 51, 111
hair, 60, 68, 69, 76
health, 88
history, 89, 91
hobby, 101
homeostasis, 106
homes, 61
hominids, 84
human, vii, viii, 5, 21, 35, 40, 42, 57, 58, 59, 60, 61, 62, 63, 65, 66, 72, 74, 77, 78, 79, 80, 81, 82, 85, 88, 91, 101
human animal, viii, 88
human behavior, 88
human development, 78

human genome, 91
human perception, 21
husbandry, 45
hybridization, viii, 107, 110
hydrocarbons, ix, 108, 109, 113, 114, 115, 116, 117, 118, 119
hyperactivity, 21
hypothalamus, 88
hypothesis, 99

I

identification, 109
identity, 65, 66, 76
illumination, 55
image(s), 10, 13, 14, 15, 16, 25, 47, 48, 69, 73
imagination, 67, 83, 84, 85
imitation, 77, 84
impairments, 59
in vitro, 92
individuals, 37, 60, 80, 108, 111
infants, 83
inferences, 92, 94
inhibitor, 96
initiation, 83
injury(s), 44, 50
inositol, 90, 100
insects, ix, 100, 108, 113, 115, 116, 118
integration, 99
integrity, 99
intelligence, 84
interface, 8
interference, 31
intervention, 62, 79
intracellular calcium, 90
invertebrates, 28
ion channels, 96
ions, 95
isolation, vii, viii, 107, 109, 110, 112, 113, 114, 116, 117, 118
issues, 42

J

Jacky dragon, vii, 2, 3, 4, 5, 7, 10, 11, 13, 19, 21, 26, 32, 33, 36, 37, 39, 40, 41, 42, 43, 44, 49, 51, 52, 53, 54, 55
Japan, 10, 13, 115
jumping, 7, 40, 47, 49
juveniles, 59

K

K^+, 101
kindergarten, 78

L

labeling, 80
landscape, 48
language development, 81
latency, 27
learning, 20, 50, 52, 55
Lepidoptera, 113, 114
lesions, 50, 91
light, viii, 14, 19, 47, 87, 89, 104
lipids, 100, 104, 113, 116
living arrangements, 61
lizards, vii, 2, 3, 5, 13, 16, 26, 31, 34, 36, 38, 39, 41, 42, 43, 45, 46, 48, 49, 51, 53, 54
localization, 100
love, 61
luminescence, 20

M

machinery, 89
magazines, 63
malaria, 100
mammals, 83, 101, 103
man, 102
manipulation, 3, 6, 11
mapping, 92
masking, 27, 39, 41

mass, 13, 113
mass spectrometry, 113
materials, 58, 63, 65, 66, 67
matrix, 16
matter, 85
measurement(s), 10, 21
median, 40
mental state, 84
messengers, 94, 113
metabolism, 50
mice, 91, 97, 102, 103
microscopy, 103
mimicry, 109, 115, 118
model system, 111
models, 3, 7, 16, 36, 37, 38
modifications, 94
molecular biology, 102, 103
morphology, 2, 7, 36, 40
mosaic, 115
motivation, 52
mutant, 101, 103
mutations, 90

N

National Academy of Sciences, 46, 52, 102, 103, 104, 106
national parks, 6
natural selection, 39, 41, 53
negotiation, 81
nematode, 90
nervous system, 90
Netherlands, 81, 102
neural network, 3
neural networks, 3
neural system(s), 3
neurons, 88, 89, 94, 96, 97, 100, 102, 103, 104, 105
neuroscience, 105
neutral, 20
neutral stimulus, 20
New South Wales, 3, 45
North America, 117
nurturance, 82

O

observed behavior, 58
occlusion, 97
olfaction, 89, 99
olfactory systems, viii, 87, 92
oligomers, 105
orchid, 115
organ(s), 2, 19, 88, 100, 101
organism, 19, 21, 90
osmotic stress, 90
overlap, 80
overlay, 10

P

Panama, 48
parental care, 75
parents, 62
partial completion, 57
participants, 60, 61
pathways, 14, 16
perceptual learning, 20, 49
pests, 99
pheromone signaling, vii, viii, 87, 89, 96, 97, 100
pheromones, 88, 114, 115, 116, 118, 119
phonology, 50
photographs, 6, 10
physical features, 7
physiology, 5, 20, 101
pitch, 11, 12, 42, 53
plants, 115
playing, 58, 74, 75
pleasure, 59
point mutation, 94
police, 66
population, 46, 113
positive correlation, 59
potassium, 101
predation, 42, 43
predators, 25, 26, 28, 39, 41, 42, 43
preschool, 78
preservation, 48

pretend play, vii, viii, 57, 58, 59, 60, 63, 64, 65, 67, 70, 71, 76, 77, 78, 79, 80, 84, 85
prevention, 111
prezygotic reproductive isolation, vii, viii, 108, 109, 110, 112
principles, 40
probability, 26
project, 96, 97
propagation, 88
proteins, 89, 92, 102, 104
psychology, 3, 48, 52
psychophysics, 20
purification, 101

Q

quadriceps, 117

R

random numbers, 64
realism, 36
reality, 59
receptors, 50, 89, 90, 94, 96, 99, 100, 101, 113, 117
recognition, 6, 7, 20, 35, 36, 43, 44, 45, 46, 47, 49, 52, 91, 109, 110, 111, 113, 114, 115, 117, 118, 119
redundancy, 90
reinforcement, 115
relatives, 112
reliability, 70
repellent, 100
replication, 90
reproduction, viii, 107, 111
reptile, 54
repulsion, 118
researchers, vii, 1, 7, 37, 40, 45, 58, 60, 72, 80, 97
residues, 93, 94
resolution, 10, 11, 13, 36
resources, 20, 43
response, 3, 21, 30, 32, 34, 38, 50, 51, 52, 88, 89, 94, 97, 99, 100, 118
responsiveness, 20
rhythmicity, 49
rings, 69, 71, 72
role playing, 77
roots, 81
routines, 77, 80
Royal Society, 46, 47, 51, 53
rules, 20, 48

S

scent, 118
segregation, 89, 96
selectivity, 92, 99, 101
senses, viii, 87, 89
sensing, viii, 87, 88, 89, 90, 91, 95, 96, 103, 104, 106
sensitivity, 2, 6, 19, 20, 21, 25, 26, 28, 36, 49, 54
sensory systems, viii, 87, 90
sex, viii, 3, 49, 54, 88, 97, 108, 109, 110, 111, 112, 113, 115, 117, 118
sexual dimorphism, 3, 110
shade, 10
shape, 11, 20, 78, 89
shock, 90, 101
showing, 23, 27, 30, 37, 93
sibling, 114
side chain, 92
signal transduction, 101
signalling, 47
signals, vii, viii, 1, 2, 19, 21, 25, 26, 28, 31, 33, 34, 35, 36, 39, 41, 42, 43, 45, 47, 48, 51, 52, 53, 54, 88, 89, 96, 99, 100, 107, 108, 109
signs, 62, 65, 67, 69, 75, 76, 80, 81, 82, 83
skeletal muscle, 105
skeleton, 11, 15
skin, 13, 36
smoothing, 11
snakes, 28
social behavior, 103
social context, 75
social hierarchy, 88
social interactions, vii, 2, 3, 6, 37

social perception, 54
software, 7, 10
speciation, viii, 107, 114, 118
species, viii, 19, 20, 21, 22, 26, 36, 42, 45, 46, 48, 50, 60, 80, 81, 87, 88, 89, 90, 92, 96, 99, 100, 104, 107, 108, 109, 110, 111, 112, 113, 114, 115, 116, 117, 118, 119
speech, 62
spiders, 40
spine, 40
Spring, 85
stimulation, 88, 89, 114
stimulus, 7, 20, 21, 32, 36, 48, 97, 98
structure, 10, 35, 36, 40, 42, 43, 44, 49, 51, 52, 53, 54, 88, 89, 92, 93, 96, 99, 100, 101
style, 108
subsistence, 59
substitution, viii, 57, 60, 65, 67, 71, 72
surface area, 63
survival, 58
symbolic thought, vii, 57, 59
symmetry, 40, 50

T

Tanzania, 82
target, 100
taxa, vii, viii, 2, 21, 77, 87, 89, 91, 92, 99, 100, 110, 112
techniques, vii, 1, 7, 52, 55
technological advances, vii, 1
technology, 7, 100
telephone(s), 68, 72, 73, 74
temperature, 3, 49, 54, 105
tempo, 52
territorial, 26, 47, 109
territory, 20, 108
texture, 10, 11, 13, 36, 37
theory of mind, vii, 57, 59, 83, 85
toys, 62, 63, 69
trade, 11, 42
trade-off, 11, 42
training, 3, 64, 70
traits, 2, 6, 45, 50, 115
transduction, viii, 88, 91, 96, 103, 105
transformation, 58, 80, 83
transient receptor potential, viii, 87, 88, 100, 103, 104, 105, 106
transmission, 26, 82
treatment, 33
trial, 23
triggers, viii, 87
TRP channels, vii, viii, 88, 89, 90, 91, 92, 93, 94, 95, 96, 97, 98, 99, 100, 101, 102, 104, 105, 106

U

urban, 42, 48, 51, 53, 55
urine, 97
UV, 10, 13

V

vacuum, 77
variables, 40
variations, 11, 21, 79, 96, 116
vegetation, 6, 25, 26, 27, 28, 30, 31, 32, 33, 34, 39, 41, 42
velocity, 16, 18, 25
vertebrae, 11
vertebrates, 48, 91, 102
videos, 16
videotape, viii, 57, 60, 63, 64, 65, 67, 70, 71, 73, 74, 75, 79
vision, 46, 49, 55
visual area, 50
visual communication, vii, 3, 7, 37, 40, 51
visual field, 33
visual modality, 7
visual processing, 19
visual stimuli, vii, 1, 51
visual stimulus, 7
visual system, 5, 16, 19, 20, 21, 26, 27, 41
vocalizations, 53
vomeronasal, viii, 87, 88, 91, 92, 94, 96, 97, 100, 101, 102

W

WHO, 62
wild type, 97
wind speed(s), 32
workers, 116
wrestling, 75

Y

yeast, 90, 101